T0269318

CAMBRIDGE LIBRARY COLLECTION

Books of enduring scholarly value

Physical Sciences

From ancient times, humans have tried to understand the workings of the world around them. The roots of modern physical science go back to the very earliest mechanical devices such as levers and rollers, the mixing of paints and dyes, and the importance of the heavenly bodies in early religious observance and navigation. The physical sciences as we know them today began to emerge as independent academic subjects during the early modern period, in the work of Newton and other 'natural philosophers', and numerous sub-disciplines developed during the centuries that followed. This part of the Cambridge Library Collection is devoted to landmark publications in this area which will be of interest to historians of science concerned with individual scientists, particular discoveries, and advances in scientific method, or with the establishment and development of scientific institutions around the world.

The Common Sense of the Exact Sciences

A student of Trinity College and a member of the Cambridge Apostles, William Kingdon Clifford (1845–79) graduated as second wrangler in the mathematical tripos, became a professor of applied mathematics at University College London in 1871, and was elected a fellow of the Royal Society in 1874. The present work was begun by Clifford during a remarkably productive period of ill health, yet it remained unfinished at his death. The statistician and philosopher of science Karl Pearson (1857–1936) was invited to edit and complete the work, finally publishing it in 1885. It tackles five of the most fundamental areas of mathematics – number, space, quantity, position and motion – explaining each one in the most basic terms, as well as deriving several original results. Also demonstrating the rationale behind these five concepts, the book particularly pleased a later Cambridge mathematician, Bertrand Russell, who read it as a teenager.

Cambridge University Press has long been a pioneer in the reissuing of out-of-print titles from its own backlist, producing digital reprints of books that are still sought after by scholars and students but could not be reprinted economically using traditional technology. The Cambridge Library Collection extends this activity to a wider range of books which are still of importance to researchers and professionals, either for the source material they contain, or as landmarks in the history of their academic discipline.

Drawing from the world-renowned collections in the Cambridge University Library and other partner libraries, and guided by the advice of experts in each subject area, Cambridge University Press is using state-of-the-art scanning machines in its own Printing House to capture the content of each book selected for inclusion. The files are processed to give a consistently clear, crisp image, and the books finished to the high quality standard for which the Press is recognised around the world. The latest print-on-demand technology ensures that the books will remain available indefinitely, and that orders for single or multiple copies can quickly be supplied.

The Cambridge Library Collection brings back to life books of enduring scholarly value (including out-of-copyright works originally issued by other publishers) across a wide range of disciplines in the humanities and social sciences and in science and technology.

The Common Sense
of the Exact Sciences

WILLIAM KINGDON CLIFFORD

THE

INTERNATIONAL SCIENTIFIC SERIES

VOL. LI.

THE

COMMON SENSE

OF THE

EXACT SCIENCES

BY THE LATE

WILLIAM KINGDON CLIFFORD

WITH 100 FIGURES

'For information commences with the senses; but the whole business
terminates in works. . . . The chief cause of failure in work (especially
after natures have been diligently investigated) is the ill determination
and measurement of the forces and actions of bodies. Now the forces
and actions of bodies are circumscribed and measured, either by distances
of space, or by moments of time, or by concentration of quantity, or by
predominance of virtue ; and unless these four things have been well
and carefully weighed, we shall have sciences, fair perhaps in theory,
but in practice inefficient. The four instances which are useful in this
point of view I class under one head as Mathematical Instances and
Instances of Measurement'—*Novum Organum*, Lib. ii, Aph. xliv

LONDON

KEGAN PAUL, TRENCH, & CO., 1 PATERNOSTER SQUARE

1885

after very careful revision, and that its title should be changed to *The Common Sense of the Exact Sciences.*

Upon Clifford's death the labour of revision and completion was entrusted to Mr. R. C. Rowe, then Professor of Pure Mathematics at University College, London. That Professor Rowe fully appreciated the difficulty and at the same time the importance of the task he had undertaken is very amply evidenced by the time and care he devoted to the matter. Had he lived to complete the labour of editing, the work as a whole would have undoubtedly been better and more worthy of Clifford than it at present stands. On the sad death of Professor Rowe, in October 1884, I was requested by Messrs. Kegan Paul, Trench, & Co. to take up the task of editing, thus left incomplete. It was with no light heart, but with a grave sense of responsibility that I undertook to see through the press the labour of two men for whom I held the highest scientific admiration and personal respect. The reader will perhaps appreciate my difficulties better when I mention the exact state of the work when it came into my hands. Chapters I. and II., *Space* and *Number* ; half of Chapter III., *Quantity* (then erroneously termed Chapter IV.) ; and Chapter V., *Motion*, were in proof. With these proofs I had only some half-dozen pages of the corresponding manuscript, all the rest having un-

PREFACE.

In March 1879 Clifford died at Madeira; six years afterwards a posthumous work is for the first time placed before the public. Some explanation of this delay must be attempted in the present preface.[1]

The original work as planned by Clifford was to have been entitled *The First Principles of the Mathematical Sciences Explained to the Non-Mathematical*, and to have contained six chapters, on *Number*, *Space*, *Quantity*, *Position*, *Motion*, and *Mass* respectively. Of the projected work Clifford in the year 1875 *dictated* the chapters on Number and Space completely, the first portion of the chapter on Quantity, and somewhat later nearly the entire chapter on Motion. The first two chapters were afterwards seen by him in proof, but never finally revised. Shortly before his death he expressed a wish that the book should only be published

[1] A still more serious delay seems likely to attend the publication of the second and concluding part (*Kinetic*) of Clifford's *Elements of Dynamic*, the manuscript of which was left in a completed state. I venture to think the delay a calamity to the mathematical world.

fortunately been considered of no further use, and accordingly destroyed. How far the contents of the later proofs may have represented what Clifford dictated I have had no means of judging except from the few pages of manuscript in my possession. In revising the proofs of the first two chapters, which Clifford himself had seen, I have made as little alteration as possible, only adding an occasional foot-note where it seemed necessary. From page 65 onwards, however, I am, with three exceptions in Chapter V., responsible for all the figures in the book.

After examining the work as it was placed in my hands, and consulting Mrs. Clifford, I came to the conclusion that the chapter on Quantity had been misplaced, and that the real gaps in the work were from the middle of Chapter III. to Chapter V., and again at the end of Chapter V. As to the manner in which these gaps were to be filled I had no definite information whatever; only after my work had been completed was an early plan of Clifford's for the book discovered. It came too late to be of use, but it at least confirmed our rearrangement of the chapters.

For the latter half of Chapter III. and for the whole of Chapter IV. (pp. 116–226) I am alone responsible. Yet whatever there is in them of value I owe to Clifford; whatever is feeble or obscure is my own.

With Chapter V. my task has been by no means light. It was written at a time when Clifford was much occupied with his theory of 'Graphs,' and found it impossible to concentrate his mind on anything else : parts of it are clear and succinct, parts were such as the author would never have allowed to go to press. I felt it impossible to rewrite the whole without depriving the work of its right to be called Clifford's, and yet at the same time it was absolutely necessary to make considerable changes. Hence it is that my revision of this chapter has been much more extensive than in the case of the first two. With the result I fear many will be dissatisfied; they will, however, hardly be more conscious of its deficiencies than I am. I can but plead the conditions under which I have had to work. One word more as to this chapter. Without any notice of mass or force it seemed impossible to close a discussion on motion; something I felt must be added. I have accordingly introduced a few pages on the laws of motion. I have since found that Clifford intended to write a concluding chapter on mass. How to express the laws of motion in a form of which Clifford would have approved was indeed an insoluble riddle to me, because I was unaware of his having written anything on the subject. I have accordingly expressed, although with great hesitation, my own views on the subject;

these may be concisely described as a strong desire to
see the terms matter and force, together with the ideas
associated with them, entirely removed from scientific
terminology—to reduce, in fact, all dynamic to kinematic. I should hardly have ventured to put forward
these views had I not recently discovered that they have
(allowing for certain minor differences) the weighty
authority of Professor Mach, of Prag.[1] But since writing
these pages I have also been referred to a discourse
delivered by Clifford at the Royal Institution in 1873,
some account of which appeared in *Nature*, June 10,
1880. Therein it is stated that ' no mathematician
can give any meaning to the language about matter,
force, inertia used in current text-books of mechanics.'[2]
This fragmentary account of the discourse undoubtedly
proves that Clifford held on the categories of matter
and force as clear and original ideas as on all subjects
of which he has treated; only, alas! they have not
been preserved.

In conclusion I must thank those friends who have
been ever ready with assistance and advice. Without
their aid I could not have accomplished the little that

[1] See his recent book, *Die Mechanik in ihrer Entwickelung.* Leipzig,
1883.

[2] Mr. R. Tucker, who has kindly searched Clifford's note-books for
anything on the subject, sends me a slip of paper with the following
words in Clifford's handwriting : ' Force is not a fact at all, but an idea
embodying what is approximately the fact.'

has been done. My sole desire has been to give to the public as soon as possible another work of one whose memory will be revered by all who have felt the invigorating influence of his thought. Had this work been published as a fragment, even as many of us wished, it would never have reached those for whom Clifford had intended it. Completed by another hand, we can only hope that it will perform some, if but a small part, of the service which it would undoubtedly have fulfilled had the master lived to put it forth.

K. P.

UNIVERSITY COLLEGE, LONDON :
February 26.

CONTENTS.

CHAPTER I.

NUMBER.

CHAPTER II.

SPACE.

CHAPTER III.

QUANTITY.

CHAPTER IV.

POSITION.

CHAPTER V.

MOTION.

THE

COMMON SENSE

OF THE

EXACT SCIENCES.

——◦—

CHAPTER I.

NUMBER.

§ 1. *Number is Independent of the order of Counting.*

THE word which stands at the head of this chapter
contains six letters. In order to find out that there
are six, we count them; *n* one, *u* two, *m* three, *b* four,
e five, *r* six. In this process we have taken the letters
one by one, and have put beside them six words which
are the first six out of a series of words that we always
carry about with us, the names of numbers. After putting
these six words one to each of the letters of the word
number, we found that the last of the words was *six*; and
accordingly we called that set of letters by the name six.

If we counted the letters in the word ' chapter ' in
the same way, we should find that the last of the
numeral words thus used would be *seven*; and accordingly we say that there are seven letters.

But now a question arises. Let us suppose that the
letters of the word *number* are printed upon separate

B

small pieces of wood belonging to a box of letters ; that we put these into a bag and shake them up and bring them out, putting them down in any other order, and then count them again ; we shall still find that there are six of them. For example, if they come out in the alphabetical order *b e m n r u,* and we put to each of these one of the names of numbers that we have before used, we shall still find that the last name will be six. In the assertion that any group of things consists of six things, it is implied that the word six will be the last of the ordinal words used, in whatever order we take up this group of things to count them. That is to say, *the number of any set of things is the same in whatever order we count them.*

Upon this fact, which we have observed with regard to the particular number six, and which is true of all numbers whatever, the whole of the science of number is based. We shall now go on to examine some theorems about numbers which may be deduced from it.

§ 2. *A Sum is Independent of the order of Adding.*

Suppose that we have two groups of things ; say the letters in the word ' number,' and the letters in the word ' chapter.' We may count these groups separately, and find that they come respectively to the numbers six and seven. We may then put them all together, and we find in this case that the aggregate group which is so formed consists of thirteen letters.

Now this operation of putting the things all together may be conceived as taking place in two different ways. We may first of all take the six things and put them in a heap, and then we may add the seven things to them one by one. The process of counting, if it is performed in this order, amounts to counting seven more ordinal words after the word six. We may however take the seven things first and put them into a heap, and then add the six things one by one to them. In this case the process of counting amounts to counting six more ordinal words after the word seven.

But from what we observed before, that if we count any set of things we come to the same number in whatever order we count them, it follows that the number we arrive at, as belonging to the whole group of things, must be the same whichever of these two processes we use. This number is called the *sum* of the two numbers 6 and 7; and, as we have seen, we may arrive at it either by the first process of adding 7 to 6, or by the second process of adding 6 to 7.

The process of adding 7 to 6 is denoted by a shorthand symbol, which was first used by Leonardo da Vinci. A little Maltese cross (+) stands for the Latin *plus*, or the English *increased by*. Thus the words *six increased by seven* are written in shorthand 6 + 7. Now we have arrived at the result that *six increased by seven is the same number as seven increased by six*. To write this wholly in shorthand, we require a symbol for the words, *is the same number as*. The symbol for these is = ; it was first used by an Englishman, Robert Recorde. Our result then may be finally written in this way :—

$$6 + 7 = 7 + 6.$$

The proposition which we have written in this symbolic form states that the sum of two numbers 6 and 7 is independent of the order in which they are added together. But this remark which we have made about two particular numbers is equally true of any two numbers whatever, in consequence of our fundamental assumption that the number of things in any group is independent of the order in which we count them. For by the sum of any two numbers we mean a number which is arrived at by taking a group of things containing the first number of individuals, and adding to them one by one another group of things containing the second number of individuals; or, if we like, by taking a group of things containing the second number of individuals, and adding to them one by one the group of things containing the first number of individuals. Now, in virtue of our fundamental assumption, the results of these two operations must be the same. Thus we have a right to say, not only that $6 + 7 = 7 + 6$, but also that $5 + 13 = 13 + 5$, and so on, whatever two numbers we like to take.

This we may represent by a method which is due to Vieta, viz., by denoting each number by a letter of the alphabet. If we write a in place of the first number in either of these two cases, or in any other case, and b in place of the second number, then our formula will stand thus :—

$$a + b = b + a.$$

By means of this representation we have made a statement which is not about two numbers in particular, but about all numbers whatever. The letters a and b so used are something like the names which we give to things, for example, the name *horse*. When we say a horse has four legs, the statement will do for any particular horse whatever. It says of that particular horse that it has four legs. If we said ' a horse has as many legs as an ass,' we should not be speaking of any particular horse or of any particular ass, but of any horse whatever and of any ass whatever. Just in the same way, when we assert that $a + b = b + a$, we are not speaking of any two particular numbers, but of all numbers whatever.

We may extend this rule to more numbers than two. Suppose we add to the sum $a + b$ a third number, c, then we shall have an aggregate group of things which is formed by putting together three groups, and the number of the aggregate group is got by adding together the numbers of the three separate groups. This number, in virtue of our fundamental assumption, is the same in whatever order we add the three groups together, because it is always the same set of things that is counted. Whether we take the group of a things first, and then add the group of b things to it one by one, and then to this compound group of $a + b$ things add the group of c things one by one; or whether we take the group of c things, and add to it the group of b things, and then to the compound group of $c + b$ things add the group of a things, the sum must in both cases be the same. We may write this result in the symbolic form $a + b + c = c + b + a$, or we may state in words that *the sum of three numbers is independent of the order in which they are added together*.

This rule may be extended to the case of any number of numbers. However many groups of things we have, if we put them all together, the number of things in the resulting aggregate group may be counted in various ways. We may start with counting any one of the original groups, then we may follow it with any one of the others, following these by any one of those left, and so on. In whatever order we have taken these groups, the ultimate process is that of counting the whole aggregate group of things; and consequently the numbers that we arrive at in these different ways must all be the same.

§ 3. *A Product is Independent of the order of Multiplying.*

Now let us suppose that we take six groups of things which all contain the same number, say 5, and that we want to count the aggregate group which is made by putting all these together. We may count the six groups of five things one after another, which amounts to the same thing as adding 5 five times over to 5. Or if we like we may simply mix up the whole of the six groups, and count them without reference to their previous grouping. But it is convenient in this case to consider the six groups of five things as arranged in a particular way.

Let us suppose that all these things are dots which are made upon paper, that every group of five things is five dots arranged in a horizontal line, and that the six groups are placed vertically under one another as in the figure,

We then have the whole of the dots of these six groups arranged in the form of an oblong which contains six rows of five dots each. Under each of the five dots belonging to the top group there are five other dots belonging to the remaining groups; that is to say, we have not only six *rows* containing five dots each, but five *columns* containing six dots each. Thus the whole set of dots can be arranged in five groups of six each, just as well as in six groups of five each. The whole number of things contained in six groups of five each, is called six times five. We learn in this way therefore that six times five is the same number as five times six.

As before, the remark that we have here made about two particular numbers may be extended to the case of any two numbers whatever. If we take any number of groups of dots, containing all of them the same number of dots, and arrange these as horizontal lines one under the other, then the dots will be arranged not only in lines but in columns; and the number of dots in every column will obviously be the same as the number of groups, while the number of columns will be equal to the number of dots in each group. Consequently the number of things in a groups of b things each is equal to the number of things in b groups of a things each, no matter what the numbers a and b are.

The number of things in a groups of b things each is called a times b; and we learn in this way that a times b is equal to b times a. The number a times b is denoted by writing the two letters a and b together, a coming first; so that we may express our result in the symbolic form $ab = ba$.

Suppose now that we put together seven such compound groups arranged in the form of an oblong like that we constructed just now. They cannot now be represented on one sheet of paper, but we may suppose that instead of dots we have little cubes which can be put into an oblong box. On the floor of the box we shall have six rows of five cubes each, or five columns of six cubes each; and there will be seven such layers, one on the top of another. Upon every cube therefore which is in the bottom of the box there will be a pile of six cubes, and we shall have altogether five times six such piles. That is to say, we have five times six groups of seven cubes each, as well as seven groups of five times six cubes each. The whole number of cubes is independent of the order in which they are counted, and consequently we may say that seven times five times six is the same thing as five times six times seven.

But it is here very important to notice that when we say seven times five times six, what we mean is that seven layers have been formed, each of which contains five times six things; but when we say five times six times seven, we mean that five times six columns have been formed, each of which contains seven things. Here it is clear that in the one case we have first multiplied the last two numbers, and then multiplied the result by the first mentioned (seven times five times six = seven times thirty), while in the other case it is the first two numbers mentioned that are multiplied together, and then the third multiplied by the result (five times six times seven = thirty times seven). Now it is quite evident that when the box is full of these cubes it may be set upon any side or upon any end; and in all cases there will be a number of layers of cubes, either 5 or 6 or 7. And whatever is the number of layers of cubes, that will also be the number of cubes in each pile. Whether therefore we take seven layers containing five times six cubes each, or six layers containing seven times five cubes each, or five layers containing six times seven cubes each, it comes to exactly the same thing.

We may denote five times six by the symbol 5×6, and then we may write five times six times seven, $5 \times 6 \times 7$.

But now this form does not tell us whether we are to multiply together 6 and 7 first, and then take 5 times the result, or whether we are to multiply 5 and 6 first, and take that number of sevens. The distinction between these two operations may be pointed out by means of parentheses or brackets; thus, $5 \times (6 \times 7)$ means that the 6 and 7 must be first multiplied together and 5 times the result taken, while $(5 \times 6) \times 7$ means that we are to multiply 5 and 6 and then take the resulting number of sevens.

We may now state two facts that we have learned about multiplication.

First, that the brackets make no difference in the result, although they do make a difference in the process by which the result is attained; that is to say, $5 \times (6 \times 7) = (5 \times 6) \times 7$.

Secondly, that the product of these three numbers is independent of the order in which they are multiplied together.

The first of these statements is called the *associative* law of multiplication, and the second the *commutative* law.

Now these remarks that we have made about the result of multiplying together the particular three numbers, 5, 6, and 7, are equally applicable to any three numbers whatever.

We may always suppose a box to be made whose height, length, and breadth will hold any three numbers of cubes. In that case the whole number of cubes will clearly be independent of the position of the box; but however the box is set down it will contain a certain number of layers, each layer containing a certain number of rows, and each row containing a certain number of cubes. The whole number of cubes in the box will then be the product of these three numbers; and it will be got at by taking any two of the three numbers, multiplying them together, and then multiplying the result by the third number.

This property of any three numbers whatever may now be stated symbolically.

In the first place it is true that $a(bc) = (ab)c$; that is, it comes to the same thing whether we multiply the product of the second and third numbers by the first, or the third number by the product of the first and second.

In the next place it is true that $abc = acb = bca$, &c., and we may say that the product of any three numbers is independent of the order and of the mode of grouping in which the multiplications are performed.

We have thus made some similar statements about two numbers and three numbers respectively. This naturally suggests to us that we should inquire if corresponding statements can be made about four or five numbers, and so on.

We have arrived at these two statements by considering the whole group of things to be counted as arranged in a layer and in a box respectively. Can we go any further, and so arrange a number of boxes as to exhibit in this way the product of four numbers? It is pretty clear that we cannot.

Let us therefore now see if we can find any other sort of reason for believing that what we have seen to be true in the case of three numbers—viz., that the result of multiplying them together is independent of the order of multiplying—is also true of four or more numbers.

In the first place we will show that it is possible to interchange the order of a pair of these numbers which are next to one another in the process of multiplying, without altering the product.

Consider, for example, the product of four numbers, *abcd*. We will endeavour to show that this is the same thing as the product *acbd*. The symbol *abcd* means that we are to take *c* groups of *d* things and then *b* groups like the aggregate so formed, and then finally *a* groups of *bcd* things.

Now, by what we have already proved, *b* groups of *cd* things come to the same number as *c* groups of *bd* things. Consequently, *a* groups of *bcd* things are the same as *a* groups of *cbd* things ; that is to say, *abcd* = *acbd*.

It will be quite clear that this reasoning will hold no matter how many letters come after *d*. Suppose, for example, that we have a product of six numbers *abcdef*. This means that we are to multiply *f* by *e*, the result by *d*, then *def* by *c*, and so on.

Now in this case the product *def* simply takes the place which the number *d* had before. And *b* groups of *c* times *def* things come to the same number as *c* groups of *b* times *def* things, for this is only the product of three numbers, *b*, *c*, and *def*. Since then this result is the same in whatever order *b* and *c* are written, there can be no alteration made by multiplications coming after, that is to say if we have to multiply by ever so many more numbers after multiplying by *a*. It follows therefore that no matter how many numbers are multiplied together, we may change the places of any two of them which are close together without altering the product.

In the next place let us prove that we may change the places of any two which are not close together. For example, that *abcdef* is the same thing as *aecdbf*, where *b* and *e* have been interchanged. We may do this by first making the *e* march backwards, changing places successively with *d* and *c* and *b*, when the product is changed into *aebcdf*; and then making *b* march forwards so as to change places successively with *c* and *d*, whereby we have now got *e* into the place of *b*.

Lastly, I say that by such interchanges as these we can produce any alteration in the order that we like. Suppose for example that I want to change *abcdef* into *dcfbea*. Here I will first get *d* to the beginning; I therefore interchange it with *a*, producing *dbcaef*. Next, I must get *c* second; I do this by interchanging it with *b*, this gives *dcbaef*. I must now put *f* third by interchanging it with *b*, giving *dcfaeb*, next put *b* fourth by interchanging it with *a*, producing *dcfbea*. This is the form required. By five such interchanges at most, I can alter the order of six letters in any way I please. It has now been proved that this alteration in the order may be produced by successive interchanges of two letters which are close together. But these interchanges, as we have before shown, do not alter the product; consequently the product of six numbers in any order is equal to the product of the same six numbers in any other order; and it is easy to see how the same process will apply to any number of numbers.

But is not all this a great deal of trouble for the sake of proving what we might have guessed beforehand? It is true we might have guessed beforehand that a product was independent of the order and grouping of its factors; and we might have done good work by developing the consequences of this guess before we were quite sure that it was true. Many beautiful theorems have been guessed and widely used before they were conclusively proved; there are some even now in that state. But at some time or other the inquiry has to be undertaken, and it always clears up our ideas about the nature of the theorem, besides giving us the right to say that it is true. And this is not all; for in most cases the same mode of proof or of investigation can be applied to other subjects in such a way as to increase our knowledge. This happens with the proof we have just gone through; but at present, as we have only numbers to deal with, we can only go backwards and not forwards in its application. We have been reasoning about multiplication; let us see if the same reasoning can be applied to addition.

What we have proved amounts to this. A certain result has been got out of certain things by taking them in a definite order; and it has been shown that *if we can interchange any two consecutive things without altering the result, then we may make any change whatever in the order without altering the result.* Let us apply this to counting. The process of counting consists in taking certain things in a definite order, and applying them to our fingers one by one; the result depends on the last finger, and its name is called the number of the things so counted. We learn then that this result will be independent of the order of counting, provided only that it remains unaltered when we interchange any two consecutive things; that is, provided that two adjacent fingers can be crossed, so that each rests on the object previously under the other, without employing any new fingers or setting free any that are already employed. With this assumption we can *prove* that the number of any set of things is independent of the order of counting; a statement which, as we have seen, is the foundation of the science of number.

§ 4. *The Distributive Law.*

There is another law of multiplication which is, if possible, still more important than the two we have already considered. Here is a particular case of it: the number 5 is the sum of 2 and 3, and 4 times 5 is the sum of 4 times 2 and 4 times 3. We can make this visible by an arrangement of dots as follows :—

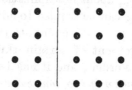

Here we have four rows of five dots each, and each row is divided into two parts, containing respectively two dots and three dots. It is clear that the whole number of dots may be counted in either of two ways; as four rows of five dots, or as four rows of two dots together with four rows of three dots. By our general principle the result is independent of the order of counting, and therefore

$$4 \times 5 = (4 \times 2) + (4 \times 3);$$

or, if we put in evidence that $5 = 2 + 3$,

$$4 (2 + 3) = (4 \times 2) + (4 \times 3).$$

The process is clearly applicable to any three numbers whatever, and not only to the particular numbers 4, 2, 3. We may construct an oblong containing a rows of $b + c$ dots ; and this may be divided by a vertical line into a rows of b dots and a rows of c dots. Counted in one way, the whole number of dots is $a(b + c)$;

counted in another way, it is $ab + ac$. Hence we must always have

$$a\,(b + c) = ab + ac.$$

This is the *first form* of the distributive law.

Now the result of multiplication is independent of the order of the factors, and therefore

$$a\,(b + c) = (b + c)\,a,$$
$$ab = ba,$$
$$ac = ca\,;$$

so that our equation may be written in the form

$$(b + c)\,a = ba + ca.$$

This is called the *second form* of the distributive law. Using the numbers of our previous example, we say that since 5 is the sum of 2 and 3, 5 times 4 is the sum of 2 times 4 and 3 times 4. This form may be arrived at independently and very simply as follows. We know that 2 things and 3 things make 5 things, whatever the things are ; let each of these things be a group of 4 things ; then 2 fours and 3 fours make 5 fours, or

$$(2 \times 4) + (3 \times 4) = 5 \times 4.$$

The rule may now be extended. It is clear that our oblong may be divided by vertical lines into more parts than two, and that the same reasoning will apply. This

figure, for example, makes visible the fact that just as 2 and 3 and 4 make 9, so 4 times 2, and 4 times 3, and 4 times 4 make 4 times 9. Or generally—

$$a \, (b + c + d) = ab + ac + ad,$$
$$(b + c + d) \, a = ba + ca + da \, ;$$

and the same reasoning applies to the addition of any
number of numbers and their subsequent multiplication.

§ 5. *On Powers.*

When a number is multiplied by itself it is said to
be squared.   The reason of this is that if we arrange a
number of lines of equally distant dots in an oblong, the
number of lines being equal to the number of dots in
each line, the oblong will become a square.

If the square of a number is multiplied by the
number itself, the number is said to be *cubed* ; because if
we can fill a box with cubes whose height, length, and
breadth are all equal to one another, the shape of the
box will be itself a cube.

If we multiply together four numbers which are all
equal, we get what is called the fourth power of any one
of them ; thus if we multiply 4 3's we get 81, if we
multiply 4 2's we get 16.

If we multiply together any number of equal numbers, we get in the same way a power of one of them
which is called its fifth, or sixth, or seventh power, and
so on, according to the number of numbers multiplied
together.

Here is a table of the powers of 2 and 3 :—

Index	1	2	3	4	5	6	7	8
Powers of 2 . . .	2	4	8	16	32	64	128	256
„ 3 . . .	3	9	27	81	243	729	2187	6561

The number of equal factors multiplied together is
called the *index*, and it is written as a small figure
above the line on the right-hand side of the number
whose power is thus expressed.   To write in shorthand

the statement that if you multiply seven threes together
you get 2187, it is only needful to put down :—

$$3^7 = 2187.$$

It is to be observed that every number is its own
first power; thus $2^1 = 2$, $3^1 = 3$, and in general $a^1 = a$.

§ 6. *Square of $a + 1$.*

We may illustrate the properties of square numbers
by means of a common arithmetical puzzle, in which
one person tells the number another has thought of by
means of the result of a round of calculations performed with it.

Think of a number . . . . say 3
Square it 9
Add 1 to the original number . . . 4
Square that 16
Take the difference of the two squares . 7

This last is always an odd number, and the number
thought of is what we may call the *less half* of it; viz.,
it is the half of the even number next below it. Thus,
the result being given as 7, we know that the number
thought of was the half of 6, or 3.

We will now proceed to prove this rule. Suppose
that the square of 5 is given us, in the form of twenty-
five dots arranged in a square, how are we to form the
square of 6 from it? We may add five dots on the
right, and then five dots along the bottom, and then
one dot extra in the corner. That is, to get the square
of 6 from the square of 5, we must add one more than
twice 5 to it. Accordingly—

$$36 = 25 + 10 + 1.$$

c

And, conversely, the number 5 is the less half of the difference between its square and the square of 6.

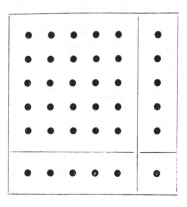

The form of this reasoning shows that it holds good for any number whatever. Having given a square of dots, we can make it into a square having one more dot in each side by adding a column of dots on the right, a row of dots at the bottom, and one more dot in the corner. That is, we must add one more than twice the number of dots in a side of the original square. If, therefore, this number is given to us, we have only to take one from it and divide by 2, to have the number of dots in the side of the original square.

We will now write down this result in shorthand. Let a be the original number; then $a+1$ is the number next above it; and what we want to say is that the square of $a+1$, that is $(a+1)^2$, is got from the square of a, which is a^2, by adding to it one more than twice a, that is $2a+1$. Thus the shorthand expression is

$$(a + 1)^2 = a^2 + 2a + 1.$$

This theorem is a particular case of a more general one, which enables us to find the square of the sum of any two numbers in terms of the squares of the two numbers and their product. We will first illustrate this by means of the square of 5, which is the sum of 2 and 3.

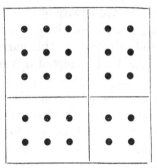

The square of twenty-five dots is here divided into two squares and two oblongs. The squares are respectively the squares of 3 and 2, and each oblong is the product of 3 and 2. In order to make the square of 3 into the square of $3+2$, we must add two columns on the right, two rows at the bottom, and then the square of 2 in the corner. And in fact, $25 = 9 + 2 \times 6 + 4$.

§ 7. On Powers of $a+b$.

To generalise this, suppose that we have a square with a dots in each side, and we want to increase it to a square with $a+b$ dots in each side. We must add b columns on the right, b rows at the bottom, and then the square of b in the corner. But each column and each row contains a dots. Hence what we have to add is twice ab together with b^2, or in shorthand :—

$$(a + b)^2 = a^2 + 2ab + b^2.$$

The theorem we previously arrived at may be got from this by making $b=1$.

Now this is quite completely and satisfactorily proved; nevertheless we are going to prove it again in another way. The reason is that we want to extend the proposition still further; we want to find an expression not only for the square of $(a+b)$, but for any other power of it, in terms of the powers and products of powers of a and b. And for this purpose the mode of proof we have hitherto adopted is unsuitable. We might, it is true, find the cube of $a+b$ by adding the proper pieces to the cube of a; but this would be somewhat cumbrous, while for higher powers no such representation can be used. The proof to which we now proceed depends on the distributive law of multiplication.

According to this law, in fact, we have

$$(a + b)^2 = (a + b)(a + b) = a(a + b) + b(a + b)$$
$$= aa + ab + ba + bb$$
$$= a^2 + 2ab + b^2.$$

It will be instructive to write out this shorthand at length. The square of the sum of two numbers means that sum multiplied by itself. But this product is the first number multiplied by the sum together with the second number multiplied by the sum. Now the first number multiplied by the sum is the same as the first number multiplied by itself together with the first number multiplied by the second number. And the second number multiplied by the sum is the same as the second number multiplied by the first number together with the second number multiplied by itself. Putting all these together, we find that the square of the sum is equal to the sum of the squares of the two numbers together with twice their product.

Two things may be observed on this comparison. First, how very much the shorthand expression gains in clearness from its brevity. Secondly, that it is only shorthand for something which is just straightforward common sense and nothing else. We may always depend upon it that algebra, which cannot be translated into good English and sound common sense, is bad algebra.

But now let us put this process into a graphical shape which will enable us to extend it. We start with two numbers, a and b, and we are to multiply each of them by a and also by b, and to add all the results.

Let us put in each case the result of multiplying by a to the left, and the result of multiplying by b to the right, under the number multiplied. The process is then shown in the figure.

If we now want to multiply this by $a + b$ again, so as to make $(a + b)^3$, we must multiply each part of the lower line by a, and also by b, and add all the results, thus :—

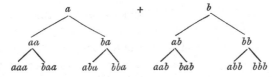

Here we have eight terms in the result. The first and last are a^3 and b^3 respectively. Of the remaining six, three are baa, aba, aab, containing two a's and one b, and therefore each equal to a^2b; and three are bba, bab, abb, containing one a and two b's, and therefore each equal to ab^2. Thus we have :—

$$(a + b)^3 = a^3 + 3a^2b + 3ab^2 + b^3.$$

For example, $11^3 = 1331$. Here $a = 10$, $b = 1$, and

$$(10+1)^3 = 10^3 + 3 \times 10^2 + 3 \times 10 + 1,$$

for it is clear that any power of 1 is 1.

We shall carry this process one step further, before making remarks which will enable us to dispense with it.

In this case there are sixteen terms, the first and last being a^4 and b^4 respectively. Of the rest, some have three a's and one b, some two a's and two b's, and some one a and three b's. There are four of the first kind, since the b may come first, second, third, or fourth; so also there are four of the third kind, for the a occurs in each of the same four places; the remaining six are of the second kind. Thus we find that,

$$(a+b)^4 = a^4 + 4a^3b + 6a^2b^2 + 4ab^3 + b^4.$$

We might go on with this process as long as we liked, and we should get continually larger and larger trees. But it is easy to see that the process of classifying and counting the terms in the last line would become very troublesome. Let us then try to save that trouble by making some remarks upon the process.

If we go down the tree last figured, from a to $abaa$, we shall find that the term

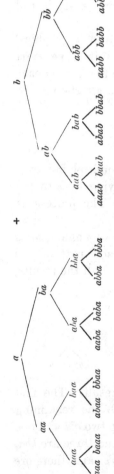

abaa is built up from right to left as we descend. The *a* that we begin with is the last letter of *abaa*; then in descending we move to the right, and put another *a* before it; then we move to the left and put *b* before that; lastly we move to the right and put in the first *a*. From this there are two conclusions to be drawn.

First, *the terms at the end are all different*; for any divergence in the path by which we descend the tree makes a difference in some letter of the result.

Secondly, *every possible arrangement of four letters which are either a's or b's is produced*. For if any such arrangement be written down, say *abab*, we have only to read it backwards, making *a* mean ' turn to the left ' and *b* ' turn to the right,' and it will indicate the path by which we must descend the tree to find that arrangement at the end.

We may put these two remarks into one by saying that *every such possible arrangement is produced once and once only*.

Now the problem before us was to count the number of terms which have a certain number of *b*'s in them. By the remark just made we have shown that this is the same thing as to count the number of possible arrangements having that number of *b*'s.

Consider for example the terms containing one *b*. When there are three letters to each term, the number of possible arrangements is 3, for the *b* may be first, second, or third, *baa, aba, aab*. So when there are four letters the number is 4, for the *b* may be first, second, third, or fourth; *baaa, abaa, aaba, aaab*. And generally it is clear that whatever be the number of letters in each term, that is also the number of places in which the *b* can stand. Or, to state the same thing in shorthand, if n be the number of letters, there are n terms containing one b. And then, of course, there are n terms containing one a and all the rest b's.

And these are the terms which come at the beginning and end of the nth power of $a+b$; viz. we must have $(a + b)^n = a^n + na^{n-1}b + $ other terms $+ nab^{n-1} + b^n$.

The meaning of this shorthand is that we have n $(a+b)$'s multiplied together, and that the result of that multiplying is the sum of several numbers, four of which we have written down. The first is the product of n a's multiplied together, or a^n; the next is n times the product of b by $(n-1)$ a's, namely, $na^{n-1}b$. The last but one is n times the product of a by $(n-1)$ b's, namely, nab^{n-1}; and the last is the product of n b's multiplied together, which is written b^n.

The problem that remains is to fill up this statement by finding what the ' other terms ' are, containing each more than one a and more than one b.

§ 8. *On the Number of Arrangements of a Group of Letters.*

This problem belongs to a very useful branch of applied arithmetic called the theory of ' permutations and combinations,' or of arrangement and selection. The theory tells us how many arrangements may be made with a given set of things, and how many selections can be made from them. One of these questions is made to depend on the other, so that there is an advantage in counting the number of arrangements first.

With two letters there are clearly two arrangements, ab and ba. With three letters there are these six :—

$$abc, \ acb, \ bca, \ bac, \ cab, \ cba,$$

namely, two with *a* at the beginning, two with *b* at the beginning, and two with *c* at the beginning; three times two. It would not be much trouble to write down all the arrangements that can be made with four letters *abcd*. But we may count the number of them without taking that trouble; for if we write *d* before each of the six arrangements of *abc*, we shall have six arrangements of the four letters beginning with *d*, and these are clearly all the arrangements which can begin with *d*. Similarly, there must be six beginning with *a*, six beginning with *b*, and six beginning with *c*; in all, four times six, or *twenty-four*.

Let us put these results together:

With two letters, number of arrangements is two = 2
  ,, three ,,   three times two . . = 6
  ,, four   ,,   four times three times two = 24

Here we have at once a rule suggested. *To find the number of arrangements which can be made with a given group of letters, multiply together the numbers two, three, four, &c., up to the number of letters in the group.* We have found this rule to be right for two, three, and four letters; is it right for any number whatever of letters?

We will consider the next case of five letters, and deal with it by a method which is applicable to all cases. Any one of the five letters may be placed first; there are then five ways of disposing of the first place. For each of these ways there are four ways of disposing of the second place; namely, any one of the remaining four letters may be put second. This makes five times four ways of disposing of the first two places. For each of these there are three ways of disposing of the third place, for any one of the remaining three letters may be put third. This makes five times four times three ways of disposing of the first three places. For each of these there are two ways of disposing of the last two places; in all, five times four times three times two, or 120 ways of arranging the five letters.

Now this method of counting the arrangements will clearly do for any number whatever of letters; so that our rule must be right for all numbers.

We may state it in shorthand thus : the number of arrangements of n letters is $1 \times 2 \times 3 \times \dots \times n$; or putting dots instead of the sign of multiplication, it is $1.2.3 \dots n$. The 1 which begins is of course not wanted for the multiplication, but it is put in to include the extreme case of there being only one letter, in which case, of course, there is only one arrangement.

The product $1.2.3 \dots n$, or, as we may say, the product of the first n natural numbers, occurs very often in the exact sciences. It has therefore been found convenient to have a special short sign for it, just as a parliamentary reporter has a special sign for 'the remarks which the Honourable Member has thought fit to make.' Different mathematicians, however, have used different symbols for it. The symbol $\lfloor n$ is very much used in England, but it is difficult to print. Some continental writers have used a note of admiration, thus, n ! Of this it has been truly remarked that it has the air of pretending that you never saw it before. I myself prefer a symbol which has the weighty authority of Gauss, namely a Greek Π (Pi), which may be taken as short for *product* if we like, thus, Πn. We may now state that—

$\Pi 1 = 1$, $\Pi 2 = 2$, $\Pi 3 = 6$, $\Pi 4 = 24$, $\Pi 5 = 120$, $\Pi 6 = 720$, and generally that

$$\Pi (n + 1) = (n + 1) \Pi n,$$

for the product of the first $n + 1$ numbers is equal to the product of the first n numbers multiplied by $n + 1$.

§ 9. *On a Theorem concerning any Power of $a + b$.*

We will now apply this rule to the problem of counting the terms in $(a+b)^n$; and for clearness' sake, as usual, we will begin with a particular case, namely the case in which $n = 5$. We know that here there is one term whose factors are all a's, and one whose factors are all b's; five terms which are the product of four a's by one b, and five which are the product of one a and four b's. It remains only to count the number of terms made by multiplying three a's by two b's, which is naturally equal to the number made by multiplying two a's by three b's. The question is, therefore, *how many different arrangements can be made with three a's and two b's?*

Here the three a's are all alike, and the two b's are alike. To solve the problem we shall have to think of them as different; let us therefore replace them for the present by capital letters and small ones. How many different arrangements can be made with three capital letters A B C and two small ones *de*?

In this question the capital letters are to be considered as equivalent to each other, and the small letters as equivalent to each other; so that the arrangement A B C *d e* counts for the *same* arrangement as C A B *e d*. Every arrangement of capitals and smalls is one of a group of $6 \times 2 = 12$ equivalent arrangements; for the 3 capitals may be arranged among one another in $\Pi 3, = 6$ ways, and the 2 smalls may be arranged in $\Pi 2, = 2$ ways. Now it is clear that by taking all the arrangements in respect of capital and small letters, and then permuting the capitals among themselves and the small letters among themselves, we shall get the whole number of arrangements of the five letters A B C $d e$; namely $\Pi5$ or 120. But since each arrangement in respect of capitals and smalls is here repeated twelve times, and since 12 goes into 120 *ten* times exactly, it appears that the number we require is ten. Or the number of arrangements of three a's and two b's is $\Pi5$ divided by $\Pi3$ and $\Pi2$.

The arrangements are in fact—

$$bbaaa,\ babaa,\ baaba,\ baaab$$
$$abbaa,\ ababa,\ abaab$$
$$aabba,\ aabab$$
$$aaabb$$

The first line has a b at the beginning, and there are *four* positions for the second b; the next line has a b in the second place, and there are *three* new positions for the other b, and so on. We might of course have arrived at the number of arrangements in this particular case by the far simpler process of direct counting, which we have used as a verification; but the advantage of our longer process is that it will give us a general formula applicable to all cases whatever.

Let us stop to put on record the result just obtained; viz. we have found that

$$(a + b)^5 = a^5 + 5a^4b + 10a^3b^2 + 10a^2b^3 + 5ab^4 + b^5.$$

Observe that $1 + 5 + 10 + 10 + 5 + 1 = 32$, that is, we have accounted for the whole of the 32 terms which would be in the last line of the tree appropriate to this case.

We may now go on to the solution of our general problem. Suppose that p is the number of a's and q is the number of b's which are multiplied together in a certain term; we want to find the number of possible arrangements of these p a's and q b's. Let us replace them for the moment by p capital letters and q small ones, making $p + q$ letters altogether. Then any arrangement of these in respect of capital letters and small ones is one of a group of equivalent arrangements got by permuting the capitals among themselves and the small letters among themselves. Now by permuting the capital letters we can make Πp arrangements, and by permuting the small letters Πq arrangements. Hence every arrangement in respect of capitals and smalls is one of a group of $\Pi p \times \Pi q$ equivalent arrangements. Now the whole number of arrangements of the $p + q$ letters is $\Pi(p+q)$; and, as we have seen, every arrangement in respect of capitals and smalls is here repeated $\Pi p \times \Pi q$ times. Consequently the number we are in search of is got by dividing $\Pi(p+q)$ by $\Pi p \times \Pi q$. This is written in the form of a fraction, thus:—

$$\frac{\Pi(p+q)}{\Pi p \cdot \Pi q},$$

although it is not a fraction, for the denominator always divides the numerator exactly. In fact, it would be absurd to talk about half a quarter of a way of arranging letters.

We have arrived then at this result, that *the number of ways of arranging p a's and q b's is*

$$\frac{\Pi(p+q)}{\Pi p \cdot \Pi q}.$$

This is also (otherwise expressed) the number of ways of dividing $p+q$ places into p of one sort and q of another; or again, it is the number of ways of selecting p things out of $p+q$ things.

Applying this now to the expression of $(a+b)^n$, we find that each of our other terms is of the form

$$\frac{\Pi n}{\Pi p \cdot \Pi q}\, a^p b^q,$$

where $p+q=n$; and that we shall get them all by giving to q successively the values 1, 2, 3, &c., and to p the values got by subtracting these from n. For example, we shall find that

$$(a+b)^6 = a^6 + 6a^5 b + \frac{\Pi 6}{\Pi 4 \cdot \Pi 2} a^4 b^2 + \frac{\Pi 6}{\Pi 3 \cdot \Pi 3}\, a^3 b^3$$
$$+ \frac{\Pi 6}{\Pi 2 \cdot \Pi 4}\, a^2 b^4 + 6ab^5 + b^6.$$

The calculation of the numbers may be considerably shortened. Thus we have to divide $1.2.3.4.5.6$ by $1.2.3.4$; the result is of course 5.6. This has to be further divided by 2, so that we finally get 5.3 or 15. Similarly, to calculate

$$\frac{\Pi 6}{\Pi 3 \cdot \Pi 3},$$

we have only to divide $4.5.6$ by $1.2.3$ or 6, and we get simply 4.5 or 20.

To write down our expression for $(a+b)^n$ we require another piece of shorthand. We have seen that it consists of a number of terms which are all of the form

$$\frac{\Pi n}{\Pi p \cdot \Pi q}\, a^p b^q,$$

but which differ from one another in having for p and q different pairs of numbers whose sum is n. Now just as we used the Greek letter Π for a product so we use the Greek letter Σ (Sigma) for a sum. Namely, the sum of all such terms will be written down thus:—

$$\Sigma \frac{\Pi n}{\Pi p \cdot \Pi q} a^p b^q, \qquad [p + q = n].$$

Now we may very reasonably include the two extreme terms a^n and b^n in the general shape of these terms. For suppose we made $p = n$ and $q = 0$, the corresponding term would be:—

$$\frac{\Pi n}{\Pi n \cdot \Pi o} a\, b^0,$$

and this will be simply a if $\Pi 0 = 1$ and $b^0 = 1$. Of course there is no sense in 'the product of the first no numbers'; but if we consider the rule

$$\Pi (n + 1) = (n + 1) \Pi n,$$

which holds good when n is any number, to be also true when n stands for nothing, and consequently $n + 1 = 1$, it then becomes

$$\Pi 1 = \Pi 0,$$

and we have already seen reason to make $\Pi 1$ mean 1. Next if we say that b^q means the result of multiplying 1 by b q times, then b^0 must mean the result of multiplying 1 by b no times, that is, of not multiplying it at all; and this result is 1.

Making then these conventional interpretations, we may say that

$$(a + b)^n = \Sigma \frac{\Pi n}{\Pi p \cdot \Pi q} a^p b^q, \qquad [p + q = n],$$

it being understood that p is to take all values from n down to 0, and q all values from 0 up to n.

This result is called the *Binomial Theorem*, and was originally given by Sir Isaac Newton. An expression containing *two* terms, like $a+b$, is sometimes called *binomial*; and the name *Binomial Theorem* is an abbreviation for *theorem concerning any power of a binomial expression.*

§ 10. *On Operations which appear to be without Meaning.*

We have so far considered the operations by which, when two numbers are given, two others can be determined from them.

First, we can add the two numbers together and get their sum.

Secondly, we can multiply the two numbers together and get their product.

To the questions what is the sum of these two numbers, and what is the product of these two numbers, there is always an answer. But we shall now consider questions to which there is not always an answer.

Suppose that I ask what number added to 3 will produce 7. I know, of course, that the answer to this is 4, and the operation of getting 4 is called subtracting 3 from 7, and we denote it by a sign and write it

$$7-3=4.$$

But if I ask, what number added to 7 will make 3, although this question seems good English when expressed in words, yet there is no answer to it; and if I write down in symbols the expression $3-7$, I am asking a question to which there is no answer.

There is then an essential difference between adding and subtracting, for two numbers always have a sum.

If I write down the expression $3+4$, I can use it as meaning something, because I know that there is a number which is denoted by that expression. But if I write down the expression $3-7$, and then speak of it as meaning something, I shall be talking nonsense, because I shall have put together symbols the realities corresponding to which will not go together. To the question, what is the result when one number is taken from another, there is only an answer in the case where the second number is greater than the first.

In the same way, when I multiply together two numbers I know that there is always a product, and I am therefore free to use such a symbol as 4×5, because I know that there is some number that is denoted by it. But I may now ask a question; I may say, What number is it which, being multiplied by 4, produces 20? The answer I know in this case is 5, and the operation by which I get it is called dividing 20 by 4. This is denoted again by a symbol, $20 \div 4 = 5$.

But suppose I say divide 21 by 4. To this there is no answer. There is no number in the sense in which we are at present using the word—that is to say, there is no whole number—which being multiplied by 4 will produce 21 : and if I take the expression $21 \div 4$, and speak of it as meaning something, I shall be talking nonsense, because I shall have put together symbols whose realities will not go together.

The things that we have observed here will occur again and again in mathematics : for every operation that we can invent amounts to asking a question, and this question may or may not have an answer according to circumstances.

If we write down the symbols for the answer to the question in any of those cases where there is no answer and then speak of them as if they meant something, we shall talk nonsense. But this nonsense is not to be thrown away as useless rubbish. We have learned by

D

very long and varied experience that nothing is more valuable than the nonsense which we get in this way; only it is to be recognised as nonsense, and by means of that recognition made into sense.

We turn the nonsense into sense by giving a new meaning to the words or symbols which shall enable the question to have an answer that previously had no answer.

Let us now consider in particular what meaning we can give to our symbols so as to make sense out of the at present nonsensical expression, $3-7$.

§ 11. *Steps.*

The operation of adding 3 to 5 is written $5+3$, and the result is 8. We may here regard the $+3$ as a way of stepping from 5 to 8, and the symbol $+3$ may be read in words, *step forward three.*

In the same way, if we subtract 3 from 5 and get 2, we write the process symbolically $5-3=2$, and the symbol -3 may be regarded as a step from 5 to 2. If the former step was forward this is backward, and we may accordingly read -3 in words, *step backwards three.*

A step is always supposed to be taken from a number which is large enough to make sense of the result. This restriction does not affect *steps forward,* because from any number we can step forward as far as we like; but backward a step can only be taken from numbers which are larger than the step itself.

The next thing we have to observe about steps is that when two steps are taken in succession from any number, it does not matter which of them comes first. If the two steps are taken in the same direction this is clear enough. $+3+4$, meaning step forward 3 and

then step forward 4, directs us to step forward by
the number which is the sum of the numbers in the
two steps; and in the same way $-3-4$ directs us to
step backward the sum of 3 and 4, that is 7.

If the steps are in opposite directions, as, for
example, $+3-7$, we have to step forward 3 and
then backward 7, and the result is that we must step
backwards 4. But the same result would have been
attained if we first stepped backward 7 and then
forward 3. The result, in fact, is always a step which
is in the direction of the greater of the two steps, and
is in magnitude equal to their difference.

We thus see that when two steps are taken in succession they are equivalent to one step, which is independent of the order in which they are taken.

We have now supplied a new meaning for our
symbols, which makes sense and not nonsense out of
the symbol $3-7$. The 3 must be taken to mean $+3$,
that is, step forward 3; the -7 must be taken to mean
step backward 7, and the whole expression no longer
means take 7 from 3, but add 3 to and then subtract
7 from any number which is large enough to make
sense of the result. And accordingly we find that the
result of this operation is -4, or, as we may write it,
$+3-7 = -4$.

From this it follows by a mode of proof precisely
analogous to that which we used in the case of multiplication, that any number of steps taken in succession
have a resultant which is independent of the order in
which they are taken, and we may regàrd this rule as
an extension of the rule already proved for the addition
of numbers.

A step may be multiplied or taken a given number
of times, for example, $2(-3) = -6$; that is to say,

if two backward steps of 3 be possible, they are equivalent to a step backwards of 6.

In this operation of multiplying a step it is clear that what we do is to multiply the number which is stepped, and to retain the character of the step. On multiplying a step forwards we still have a step forwards, and on multiplying a step backwards we still have a step backwards.

This multiplying may be regarded as an operation by which we change one step into another. Thus in the example we have just considered the multiplier 2 changes the step backwards 3 into the step backwards 6. But this operation, as we have observed, will only change a step into another of the same kind, and the question naturally presents itself, Is it possible to find an operation which shall change a step into one of a different kind? Such an operation we should naturally call reversal. We should say that a step forwards is reversed, when it is made into a step backwards; and a step backwards is reversed when it is made into a step forwards.

If we denote the operation of reversal by the letter r, we can, by combining this with a multiplication, change -3 into $+6$, a step backwards 3 into a step forwards 6; viz. we should have the expression $r2(-3) = +6$. Now the operation, which is performed on one step to change it into another, may be of two kinds: either it keeps a step in the direction which it originally had, or it reverses it. If to make things symmetrical we insert the letter k when a step is kept in its original direction, we may write the equation $k2(-3) = -6$ to express the operation of simply multiplying.

Of course it is possible to perform on any given step a succession of these operations. If I take the step +4, treble it, and reverse it, I get −12. If I double this and keep it, I get − 24, and this may be written, $k2(r3)(+4) = -24$. But this is equal to $r6(+4)$, which tells us that the two successive operations which we have performed on this step, trebling and reversing it, doubling and keeping it, are equivalent to the single operation of multiplying by 6 and reversing it. It is clear also that whatever step we had taken the two first operations performed successively are always equivalent to the third, and we may thus write the equation $k2(r3) = r6$.

Suppose however we take another step and treble it and reverse it, and then double it and reverse it again ; we should have the result of multiplying it by six and keeping its direction unchanged.

This may be written $r2(r3) = k \cdot 6$.

If we compare the last two formulæ with those which we previously obtained, viz. $k2(-3) = -6$ and $r2(-3) = +6$, we shall see that the two sets are alike except that in the one last obtained k and r are written instead of + and − respectively.

The two sets however express entirely different things. Thus, taking the second formulæ of either set on the one hand, the statement is, Double and reverse the step backward 3, and you have a step forward 6 ; on the other hand, Treble and reverse and then double and reverse any step whatever, and you have the effect of *sextupling* and keeping the step. We shall find that this analogy holds good in general, that is, if we write down the effect of any number of successive operations performed upon a step, there will always be a corresponding statement in which this stepping is replaced by an operation ; or we may say, any operation which converts one step into another will also convert one operation into another where the converted operation is a multiplying by the number expressing the step and a keeping or reversing according as the step is forward or backward.

§ 12. *Extension of the Meaning of Symbols.*

We now proceed to do something which must apparently introduce the greatest confusion, but which, on the other hand, increases enormously our powers.

Having two things which we have so far quite rightly denoted by different symbols, and finding that we arrive at results which are uniform and precisely similar to one another except that in one of them one set of symbols is used, in the other another set, we alter the meaning of our symbols so as to see only one set instead of two. We make the symbols + and − mean for the future what we have here meant by k and r, viz. keep and reverse. We give them these meanings in addition to their former meanings, and leave it to the context to show which is the right meaning in any particular case. Thus, in the equation $(-2)(-3) = +6$ there are two possible meanings ; the -3 and $+6$, may both mean steps, in this case the statement is : Double and reverse the step backwards of 3 and you get the step forward 6. But the -3 and the $+6$ may also mean not steps but operations, and in this case the meaning is triple and reverse and then double and reverse any step whatever, and you get the same result as if you had sextupled and kept the step.

Let us now see what the reason is for saying that these two meanings can always exist together. Let us first of all take the second meaning, and frame a rule for finding the result of any number of successive operations.

First, the number which is the multiplier in the result must clearly be the product of all the numbers in the successive operations.

Next, every pair of reversals cancel one another, so that, if there is an even number of them, the result must be an operation of *retaining*.

This then is the rule : Multiply together the numbers in the several operations, prefixing to them + if there is an even number of *minus* or reversing operations, prefixing — if there is an odd number.

In the next place, suppose that many successive operations are performed upon a step. The number in the resulting step will clearly be the product of all the numbers in the operations and in the original step.

If there is an even number of reversing operations, the resulting step will be of the same kind as the original one; if an odd number, of the opposite kind. Now let us suppose that the original step were a step backwards; then if there is an even number of reversing operations, the resulting step will also be a step backwards. But in this case the number of (−) signs, reckoned independently of their meaning, will be odd; and so the rule coincides with the previous one.

If an odd number of reversing operations is performed on a negative step, the result is a positive step. But here the whole number of (−) signs, irrespective of their meaning, is an even number; and the result again agrees with the previous one.

In all cases therefore by using the same symbols to mean either a 'forward' and a 'backward' step respectively, or 'keep' and 'reverse' respectively, we shall be able to give to every expression two interpretations, and neither of these will ever be untrue.

In the process of examining this statement we have

shown by the way that the result of any number of successive operations on a step is independent of the order of them. For it is always a step whose magnitude is the product of the numbers in the original step and in the operations, and whose character is determined by the number of reversals.

§ 13. *Addition and Multiplication of Operations.*

We may now go on to find a rule which connects together the multiplication and the addition of steps.

If I multiply separately the steps $+3$ and -7 by 4, and then take the resultant of the two steps which I so obtain, I shall get the same thing as if I had first formed the resultant of $+3$ and -7, and then multiplied it by 4. In fact, $+12 - 28 = -16$, which is $4(-4)$. This is true in general, and it obviously amounts to the original rule that a set of things comes to the same number in whatever order we count them. Only that now some of the counting has to be done backwards and some again forwards.

But now, besides adding together steps, we may also in a certain sense add together operations. It seems natural to assume at once that by adding together $+3$ and -7 regarded as operations, we must needs get the operation -4. It is very important not to assume anything without proof, and still more important not to use words without attaching a definite meaning to them.

The meaning is this. If I take any step whatever, treble it without altering its character, and combine the result with the result of multiplying the original step by 7 and reversing it, then I shall get the same result as if I had multiplied the original step by 4 and reversed it. This is perfectly true, and we may see it to be true by, as it were, performing our operations in the form of steps. Suppose I take the step + 5, and want to treble it and keep its character unchanged. I can do this by taking three steps of five numbers each in the same direction (viz. the forward direction) as the original step was to be taken. Similarly, if I want to multiply it by − 7, this means that I must take 7 steps of five numbers each in the opposite or backward direction. Then finally, what I have to do is to take three steps forwards and seven steps backwards, each of these steps consisting of five numbers ; and it appears at once that the result is the same as that of taking 4 steps backwards of five numbers each.

We have thus a definition of the sum of two operations ; and it appears from the way in which we have arrived at it that this sum is independent of the order of the operations.

We may therefore now write the formulæ :—

$$a + b = b + a$$
$$a\,(b + c) = ab + ac$$
$$(a + b)c = ac + bc$$
$$ab = ba,$$

and consider the letters to signify operations performed upon steps. In virtue of the truth of these laws the whole of that reasoning which we applied to finding a power of the sum of two numbers is applicable to the finding of a power of the sum of two operations. If it did not take too much time and space, we might go through it again, giving to all the symbols their new meanings.

It is worth while, perhaps, by way of example, to explain clearly what is meant by the square of the sum of two operations.

We will take for example, $+5$ and -3.

The formula tells us that $(+5-3)^2$ is equal to $(+5)^2+(-3)^2+2(+5)(-3)$. This means that if we apply to any step twice over the sum of the operations $+5$ and -3, that is to say, if we multiply it by 5 and keep its direction, and combine with this step the result of multiplying the original step by 3 and reversing it, and then apply the same process to the result so obtained, we shall get a step which might also have been arrived at by combining together the following three steps :—

First, the original step twice multiplied by 5.

Secondly, the original step twice multiplied by 3 and twice reversed; that is to say, unaltered in direction.

Thirdly, twice the result of tripling the original step and reversing it, and then multiplying by 5 and retaining the direction.

§ 14. *Division of Operations.*

We have now seen what is meant by the multiplication of operations; let us go on to consider what sort of question is asked by *division*.

Let us take for example the symbolic statement $-3(+5) = -15$; and let us give it in the first place the meaning that to triple and reverse the step forward 5 gives the step backward 15. We may ask two questions upon this statement. First, What operation is it which, being performed on the step forwards 5, will give the step backwards 15? The answer, of course, is triple and reverse. Or we may ask this question. What step is that, which, being tripled and reversed, will give the step backward 15? The answer is, Step forwards 5. But we have only one word to describe the process by which we get the answer in these two cases. In the first case we say that we *divide* the step −15 by the step +5; in the second case we say we divide the step −15 by the operation −3.

The word *divide* thus gets two distinct meanings. But it is very important to notice that symbolically the answer is the same in the two cases, although the interpretation to be given to it is different.

The step −15 may be got in two ways; by tripling and reversing the forward step +5, or by quintupling the backward step −3. In symbols,

$$(-3)(+5) = (+5)(-3) = -15.$$

Hence the problem, *Divide* −15 by −3 may mean either of these two questions: What step is that which, being tripled and reversed, gives the step −15? Or, What operation is that which, performed on the step −3, gives the step −15? The answer to the first question is, the step +5; the answer to the second is the operation of quintupling and retaining direction, that is, the operation +5. So that although the word *divide*, as we have said, gets two distinct meanings, yet the two different results of division are expressed by the same symbol.

In general we may say that the problem, Divide the step *a* by the step *b*, means, Find the operation (if any) which will convert *b* into *a*. But the problem, Divide the step *a* by the operation *b*, means, Find the step (if any) which *b* will convert into *a*. In both cases, however, the process and the symbolic result are the same. We must divide the number of *a* by the number of *b*, and prefix to it + if the signs of *a* and *b* are alike, − if they are different.

We may also give to our original equation

$$(-3) \times (+5) = -15$$

its other meaning, in which both -3 and $+5$ are operations, and -15 is the operation which is equivalent to performing one of them after the other. In this case the problem, Divide the operation -15 by the operation -3 means, Find the operation which, being succeeded by the operation -3, will be equivalent to the operation -15. Or generally, Divide the operation a by the operation b, means, Find the operation which, being succeeded by b, will be equivalent to a.

Now it is worth noticing that the division of step by step and the division of operation by operation, have a certain likeness between them, and a common difference from the division of step by operation. Namely, the result of dividing a by b, or, as we may write it, $\frac{a}{b}$, when a and b are both steps or both operations, is an operation which converts b into a. This we may write in shorthand,

$$\frac{a}{b} \cdot b = a.$$

But when a is a step and b an operation, the result of division is a step on which the operation b must be performed to convert it into a; or, in shorthand,

$$b \cdot \frac{a}{b} = a.$$

The fact that the symbolic result is the same in the two cases may be stated thus :—

$$\frac{a}{b} \cdot b = b \cdot \frac{a}{b},$$

and in this form we see that it is a case of the commutative law. So long, then, as the commutative law is true, there is no occasion for distinguishing symbolically between the two meanings. But, as we shall see

by-and-by, there is occasion to deal with other kinds
of steps and operations in which the commutative law
does not hold; and for these a convenient notation has
been suggested by Professor Cayley. Namely, $\frac{a|}{|b}$ means
the operation which makes b into a; but $\frac{|a}{b|}$ repre-
sents that which the operation b will convert into a. So
that—

$$\frac{a|}{|b} \cdot b = a, \text{ but } b \cdot \frac{|a}{b|} = a.$$

It is however convenient to settle beforehand that whenever the symbol $\frac{a}{b}$ is used without warning it is to have
the first meaning—namely, the operation which makes
b into a.

§ 15. *General Results of our Extension of Terms.*

It will be noticed that we have hereby passed from
the consideration of mere numbers, with which we
began, to the consideration first of steps of addition or
subtraction of number from number, and then of
operations of multiplying and keeping or multiplying
and reversing, performed on these steps; and that we
have greatly widened the meaning of all the words that
we have employed.

To *addition*, which originally meant the addition of
two numbers, has been given the meaning of a combination of steps to form a resultant step equivalent in effect
to taking them in succession.

To *multiplication*, which was originally applied to
two numbers only, has been given the meaning of a
combination of operations upon steps to form a resultant
operation equivalent to their successive performance.

We have found that the same properties which characterise the addition and multiplication of numbers belong also to the addition and multiplication of steps and of operations. And it was this very fact of the similarity of properties which led us to use our old words in a new sense. We shall find that this same process is carried on in the consideration of those other subjects which lie before us ; but that the precise similarity which we have here observed in the properties of more simple and more complex operations will not in every case hold good ; so that while this gradual extension of the meaning of terms is perhaps the most powerful instrument of research which has yet been used, it is always to be employed with a caution proportionate to its importance.

CHAPTER II.

SPACE.

§ 1. *Boundaries take up no Room.*

GEOMETRY is a physical science. It deals with the
sizes and shapes and distances of things. Just as we
have studied the *number* of things by making a simple
and obvious observation, and then using this over and
over again to see where it would bring us; so we shall
study the science of the shapes and distances of things
by making one or two very simple and obvious observations, and then using these over and over again, to
see what we can get out of them.

The observations that we make are :—

First, that a thing may be moved about from one
place to another without altering its size or shape.

Secondly, that it is possible to have things of the
same shape but of different sizes.

Before we can use these observations to draw any
exact conclusions from them, it is necessary to consider
rather more precisely what they mean.

Things take up room. A table, for example, takes
up a certain part of the room where it is, and there is
another part of the room where it is not. The thing
makes a difference between these two portions of space.

Between these two there is what we call the *surface*
of the table.

We may suppose that the space all round the table

is filled with air. The surface of the table is then something just between the air and the wood, which separates them from one another, and which is neither the one nor the other.

It is a mistake to suppose that the surface of the table is a very thin piece of wood on the outside of it. We can see that this is a mistake, because any reason which led us to say so, would lead us also to say that the surface was a very thin layer of air close to the table. The surface in fact is common to the wood and to the air, and takes up itself no room whatever.[1]

Part of the surface of the table may be of one colour and part may be of another.

On the surface of this sheet of paper there is drawn a round black spot. We call the black part a circle.

FIG. 1.

It divides the surface into two parts, one where it is and one where it is not.

This circle takes up room on the surface, although the surface itself takes up no room in space. We are thus led to consider two different kinds of *room*; space-room, in which solid bodies are, and in which they move about; and surface-room, which may be regarded

[1] It is certain that however smooth a *natural* surface may *appear* to be, it could be magnified to roughness. Hence, in the case of the surface of the table and the air, it would seem probable that there is a layer in which particles of wood and air are mingled. The boundary in this case of air and table would not be what we 'see and feel' (cf. p. 48), nor would it correspond to the surface of the geometer. We are, I think, compelled to consider the surface of the geometer as an 'idea or imaginary conception,' drawn from the *apparent* (not real) boundaries of physical objects, such as the writer is describing. Strongly as I feel the ideal nature of geometrical conceptions in the exact sciences, I have thought it unadvisable to alter the text. The distinction is made by Clifford himself (*Essays*, I. pp. 306-7, 321).- -K.P.

from two different points of view. From one point of view it is the boundary between two adjacent portions of space, and takes up no space-room whatever. From the other point of view it is itself also a kind of room which may be taken up by parts of it.

These parts in turn have their boundaries.

Between the black surface of the circle and the white surface of the paper round it there is a line, the circumference of the circle. This line is neither part of the black nor part of the white, but is between the two. It divides one from the other, and takes up no surface-room at all. The line is not a very thin strip of surface, any more than the surface is a very thin layer of solid.

Anything which led us to say that this line, the boundary of the black spot, was a thin strip of black, would also lead us to say that it was a thin strip of white.

We may also divide a line into two parts. If the paper with this black circle upon it were dipped into

Fig. 2.

water so that part of the black circle were submerged, then the line surrounding it would be partly in the water and partly out.

The submerged part of the line takes up room on it. It goes a certain part of the way round the circumference. Thus we have to consider line-room as well as space-room and surface-room. The line takes up absolutely no room on the surface; it is merely the boundary between two adjacent portions of it. Still less does it take up any room in space. And yet it has a certain room of its own, which may be divided into parts, and taken up or filled by those parts.

E

These parts again have boundaries. Between the submerged portion of the circumference and the other part there are two *points*, one at each end. These points are neither in the water nor out of it. They are in the surface of the water, just as they are in the surface of the paper, and on the boundary of the black spot. Upon this line they take up absolutely no room at all.

A point is not a very small length of the line, any more than the line is a very thin strip of surface. It is a division between two parts of the line which are next one another, and it takes up no room on the line at all.

The important thing to notice is that we are not here talking of ideas or imaginary conceptions, but only making common-sense observations about matters of every-day experience.

The surface of a thing is something that we constantly observe. We can see it and feel it, and it is a mere common-sense observation to say that this surface is common to the thing itself and to the space surrounding it.

A line on a surface which separates one part of the surface from another is also a matter of every-day experience. It is not an idea got at by supposing a string to become indefinitely thin, but it is a thing given directly by observation as belonging to both portions of the surface which it divides, and as being therefore of absolutely no thickness at all. The same may be said of a point. The point which divides the part of our circumference which is in water from the part which is out of water is an observed thing. It is not an idea got at by supposing a small particle to become smaller and smaller without any limit, but it is the boundary between two adjacent parts of a line, which is the boundary between two adjacent portions of a surface, which is the boundary between two adjacent portions of space. A point is a thing which we can see and know, not an abstraction which we build up in our thoughts.

When we talk of drawing lines or points on a sheet of paper, we use the language of the draughtsman and not of the geometer. Here is a picture of a cube represented by *lines*, in the draughtsman's sense. Each of these so-called 'lines' is a black streak of printer's ink, of varying breadth, taking up a certain

Fig. 3.

amount of room on the paper. By drawing such 'lines' sufficiently close together, we might entirely cover up as large a patch of paper as we liked. Each of these streaks has a line on each side of it, separating the black surface from the white surface; these are true geometrical lines, taking up no surface-room whatever. Millions of millions of them might be marked out between the two boundaries of one of our streaks, and between every two of these there would be room for millions more.

Still, it is very convenient, in drawing geometrical figures, to represent lines by black streaks. To avoid all possible misunderstanding in this matter, we shall make a convention once for all about the sense in which a black streak is to represent a line. When the streak is vertical, or comes straight down the page, like this , the *line* represented by it is its *right-hand boundary*. In all other cases the line shall be the *upper boundary* of the streak.

So also in the case of a point. When we try to represent a point by a dot on a sheet of paper, we

E 2

make a black patch of irregular shape. The boundary
of this black patch is a line. When one point of this
boundary is higher than all the other points, that
highest point shall be the one represented by the dot.
When however several points of the boundary are at
the same height, but none higher than these, so that
the boundary has a flat piece at the top of it, then the
right-hand extremity of this flat piece shall be the
point represented by the dot.

This determination of the meaning of our figures
is of no practical use. We lay it down only that the
reader may not fall into the error of taking patches
and streaks for geometrical points and lines

§ 2. *Lengths can be Moved without Change.*

Let us now consider what is meant by the first of
our observations about space, viz., that a thing can be
moved about from one place to another without altering
its size or shape.

First as to the matter of size. We measure the size
of a thing by measuring the distances of various points
on it. For example, we should measure the size of a
table by measuring the distance from end to end, or the
distance across it, or the distance from the top to the
bottom. The measurement of distance is only possible
when we have something, say a yard measure or a piece
of tape, which we can carry about and which does not
alter its length while it is carried about. The measurement is then effected by holding this thing in the place
of the distance to be measured, and observing what
part of it coincides with this distance.

Two lengths or distances are said to be *equal* when
the same part of the measure will fit both of them.

Thus we should say that two tables are equally broad,
if we marked the breadth of one of them on a piece of
tape, and then carried the tape over to the other table
and found that its breadth came up to just the same
mark. Now the piece of tape, although convenient, is
not absolutely necessary to the finding out of this fact.
We might have turned one table up and put it on top
of the other, and so found out that the two breadths
were equal. Or we may say generally that two lengths
or distances of any kind are equal, when, one of them
being brought up close to the other, they can be made
to fit without alteration. But the tape is a thing far
more easily carried about than the table, and so in practice we should test the equality of the two breadths by
measuring both against the same piece of tape. We
find that each of them is equal to the same length of
tape ; and we assume that *two lengths which are equal to
the same length are equal to each other.* This is equivalent to saying that if our piece of tape be carried
round any closed curve and brought back to its original
position, it will not have altered in length.

How so ? Let us assume that, when not used, our
piece of tape is kept stretched out on a board, with one
end against a fixed mark on the board. Then we know
what is meant by two lengths being equal which are
both measured along the tape from that end. Now take
three tables, A, B, C, and suppose we have measured
and found that the breadth of A is equal to that of B,
and the breadth of B is equal to that of C, then we
say that the breadth of A is equal to that of C. This
means that we have marked off the breadth of A on
the tape, and then carried this length of tape to B, and
found it fit. Then we have carried the same length
from B to C, and found it fit. In saying that the

breadth of C is equal to that of A, we assert that on taking the tape from C to A, whether we go near B or not, it will be found to fit the breadth of A. That is, if we take our tape from A to B, then from B to C, and then back to A, it will still fit A if it did so at first.

These considerations lead us to a very singular conclusion. The reader will probably have observed that we have defined length or distance by means of a measure which can be carried about *without changing its length*. But how then is this property of the measure to be tested? We may carry about a yard measure in the form of a stick, to test our tape with; but all we can prove in that way is that the two things are always of the same length when they are in the same place; not that this length is unaltered.

The fact is that everything would go on quite as well if we supposed that things did change in length by mere travelling from place to place, provided that (1) different things changed equally, and (2) anything which was carried about and brought back to its original position filled the same space.[1] All that is wanted is that two things which fit in one place should also fit in another place, although brought there by different paths; unless, of course, there are other reasons to the contrary. A piece of tape and a stick which fit one another in London will also fit one another in New York, although the stick may go there across the Atlantic, and the tape *via* India and the Pacific. Of course the stick may expand from damp and the tape may shrink from dryness; such non-geometrical circumstances would have to be allowed for. But so far as the geometrical conditions alone are concerned—the

---

[1] These remarks refer to the geometrical, and not necessarily to all the physical properties of bodies.—K. P

mere carrying about and change of place—two things which fit in one place will fit in another.

Upon this fact are founded, as we have seen, the notion of length as measured, and the axiom that lengths which are equal to the same length are equal to one another.

Is it possible, however, that lengths do really change by mere moving about, without our knowing it?

Whoever likes to meditate seriously upon this question will find that it is wholly devoid of meaning. But the time employed in arriving at that conclusion will not have been altogether thrown away.

§ 3. *The Characteristics of Shape.*

We have now seen what is meant by saying that a thing can be moved about without altering its size; namely, that any length which fits a certain measure in one position will also fit that measure when both have been moved by any paths to some other position. Let us now inquire what we mean by saying that a thing can be moved about without altering its shape.

First let us observe that the shape of a thing depends only on its bounding surface, and not at all upon the inside of it. So that we may always speak of the shape of the surface, and we shall mean the same thing as if we spoke of the shape of the thing.

Fig. 4.

Let us observe then some characteristics of the surface of things. Here are a cube, a cylinder, and a sphere.

The surface of the cube has six flat sides, with edges and corners. The cylinder has two flat ends and a round surface between them; the flat ends being divided from the round part by two circular edges. The sphere has a round smooth surface all over.

We observe at once a great distinction in shape between *smooth* parts of the surface, and *edges*, and *corners*. An edge being a line on the surface is not any *part* of it, in the sense of taking up surface room; still less is a corner, which is a mere point. But still we may divide the points of the surface into those where it is smooth (like all the points of the sphere, the round and flat parts of the cylinder, and the flat sides of the cube), into points on an edge, and into corners. For convenience, let us speak of these points respectively as *smooth-points*, *edge-points*, and *corner-points*. We may also put the edges and corners together, and call them *rough-points*.

Now let us take the sphere, and put it upon a flat face of the cube (fig. 5). The two bodies will be in con-

FIG. 5.

tact at one point; that is to say, a certain point on the surface of the sphere and a certain point on the surface of the cube are made to coincide with one another and to be the same point. And these are both smooth-points. Now *we cannot move the sphere ever so little without separating these points*. If we roll it a very little way on the

face of the cube, we shall find that a different point of
the sphere is in contact with a different point of the cube.

Fig. 6.

And the same thing is true if we place the sphere in
contact with a smooth-point on the cylinder (fig. 6).

Next let us put the round part of the cylinder on
the flat face of the cube. In this case there will be
contact all along a line. At any point of this line, a
certain point on the surface of the cylinder and a
certain point on the surface of the cube have been made
to coincide with one another and to be the same point.
And these are both smooth-points. It is just as true
as before, that we cannot move one of these bodies ever
so little relatively to the other without separating the

Fig. 7.

points of their surfaces which are in contact. If we
roll the cylinder a very little way on the face of the
cube, we shall find that a different line of the cylinder
is in contact with a different line of the cube. All the
points of contact are changed.

Now put the flat end of the cylinder on the face of
the cube. These two surfaces fit throughout and make
but one surface; we have contact, not (as before) at a
point or along a line, but over a surface. Let us fix

our attention upon a particular point on the flat surface
of the cylinder and the point on the face of the cube
with which it now coincides; these two being smooth-

FIG. 8.

points. We observe again, that *it is impossible to move
one of these bodies ever so little relatively to the other
without separating these two points.*[1]

Here, however, something has happened which will
give us further instruction. We have all along sup-

FIG. 9.

posed the flat face of the cylinder to be smaller than
the flat face of the cube. When these two are in con-

---

[1] In all these cases (figs. 5-8) the relative motion spoken of must be
either motion of *translation* or of *tilting*; one body might have a *spin*
about a vertical axis without any separation of these two points. The true
distinction between the contact of smooth-points and of smooth and rough-
points seems to be this : in the former case without separating two points
there is only *one* degree of freedom—namely, spin about an axis normal to
the smooth surfaces at the points in question ; in the latter case there are
at least two (edge-point or smooth-point) and may be an infinite number of
degrees of freedom—namely, spins about two or more axes passing through
the rough-point. The reader will understand these terms better after the
chapter on Motion.—K. P.

tact, let the cylinder stand on the middle of the cube, as in fig. 8, the circle being wholly enclosed by the square. Then when we tilt the cylinder over we shall get it into the position of fig. 9. We have already observed that in this case no smooth-points which were previously in contact remain in contact. But there are two points which remain in contact; for in the tilted position a point on the circular edge of the cylinder rests on a point on the face of the cube; and these two points were in contact before. We may tilt the cylinder as much or as little as we like—provided we tilt always in the same direction, not rolling the cylinder on its edge—and these two points will remain in contact. We learn therefore that *when an edge-point is in contact with a smooth-point, it may be possible to move one of the two bodies relatively to the other without separating those two points.*

The same thing may be observed if we put the round or flat surface of the cylinder against an edge of the cube, or if we put the sphere against an edge of either of the other bodies. Holding either of them fast, we may move the other so as to keep the same two points in contact; but in order to do this, we must always tilt in the same direction.

If, however, we put a *corner* of the cube in contact with a smooth point of the cylinder, as in fig. 10, we

Fig. 10. ·

shall find that we can keep these two points in contact without any restriction on the direction of tilting. We may tilt the cube any way we like, and still keep its corner in contact with the smooth-point of the cylinder.

When we put two edge-points together, it makes a difference whether the edges are in the same direction at the point of contact or whether they cross one another. In the former case we may be able to keep the same two points in contact by tilting in a particular direction ; in the latter case we may tilt in any direction. So if a corner is in contact with an edge-point there is no restriction on the direction of tilting, and much more if a corner is in contact with a corner.

The upshot of all this is, that *in a certain sense all surfaces are of the same shape at all smooth-points* ; for when we put two smooth-points in contact, the surfaces so fit one another at those points that we cannot move one of them relatively to the other without separating the points.[1]

It is possible for two edges to fit so that we cannot move either of the bodies without separating the points in contact. For this it is necessary that one of them should be re-entrant (that is, should be a depression in the surface, not a projection), as in fig. 11 ; and here

Fig. 11.

we can see the propriety of saying that the two surfaces are of the same shape at a point where they fit in this way. The body placed in contact with the cube

[1] See, however, the footnote, p. 58.—K. P.

is formed by joining together two spheres from which
pieces have been sliced off. If only very small pieces
have been sliced off, the re-entrant edge will be very
sharp, and it will be impossible to bring the cube-edge
into contact with it (fig. 12); if nearly half of each

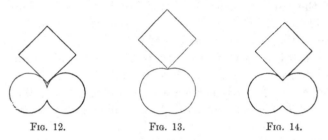

FIG. 12.   FIG. 13.   FIG. 14.

sphere has been cut off the re-entrant edge will be wide
open, and the cube will rock in it (fig. 13). There is
clearly one intermediate form in which the two edges
will just fit (fig. 14); contact at the edge will be
possible, but no rocking. Now in this case, although
one edge sticks out and the other is a dint, we may
still say that the two surfaces are of the same shape
at the edge. For if we suppose our twin-sphere
body to be made of wood, its surface is not only surface of the wood, but also surface of the surrounding
air. And that which is a dint or depression in the
wood is at the same time a projection in the air. In
just the same way, each of the projecting edges and
corners of the cube is at the same time a dint or
depression in the air. But the *surface* belongs to one
as much as the other; it knows nothing of the difference between inside and outside; elevation and depression are arbitrary terms to it. So in a thin piece of
embossed metal, elevation on one side means depression
on the other, and *vice versâ*; but it is purely arbitrary

which side we consider the *right* one. (Observe that the thin piece of metal is in no sense a representation of a *surface*; it is merely a thin solid whose two surfaces are very nearly of the same shape.)

Thus we see that the edge of wood in our cube is of the same shape as the edge of air in the twin-sphere solid; or, which is the same thing, that the two surfaces are of the same shape at the edge.

Now this twin-sphere solid is a very convenient one, because we can so modify it as to make an edge of any shape we like. Hitherto we have supposed the slices cut off to be less than half of the spheres; let us now fasten together these pieces, and so form a solid with a projecting edge, as in fig. 15. The two solids so formed, one with a re-entrant edge from the larger pieces, the other with a projecting edge from the smaller pieces, will be found always to have their edges of the same shape, or to fit one another at the edge in the sense just explained.

Fig. 15.

Now suppose that we cut our spheres very nearly in half. (Of course they must always be cut both alike, or the flat faces would not fit together.) Then when we join together the larger pieces and the smaller pieces, we shall form solids with very wide open edges. The projecting edge will be a very slight ridge, and the re-entrant one a very slight depression.

If we now go a step further, and cut our spheres actually in half, of course each of the new solids will be again a sphere; and there will be neither ridge nor

depression ; the surfaces will be smooth all over. But
we have arrived at this result by considering a project-

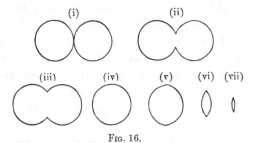

Fig. 16.

ing edge as gradually widening out until the ridge disappears, or by considering a re-entrant edge as gradually
widening out until the dint disappears. Or we may
suppose the projecting edge to go on widening out till
it becomes smooth, and then to turn into a re-entrant
edge. We might represent this process to the eye by
putting into a wheel of life a succession of pictures like
that in fig. 16, and then rapidly turning the wheel. We
should see the two spheres, at first separate, coalesce
into a single solid in (ii) and (iii), then form one sphere as
at (iv), then contract into a smaller and smaller lens at
(v), (vi), (vii). The important thing to notice is that the
single sphere at (iv) is a step in the process; or, what
is the same thing, that *a smooth-point is a particular case
of an edge-point coming between the projecting and the re-
entrant edges.* As being this particular case of the
edge-point, we say that at all smooth-points the surfaces are of the same shape.

§ 4. *The Characteristics of Surface Boundaries.*

Remarks like these that we have made about solid
bodies or portions of space may be made also about

portions of surface. Only we cannot now say that the
shape of a piece of surface depends wholly on that of
the curve which bounds it. Still the only thing that
remains for us to consider is the shape of the boundary,
because we have already discussed (so far as we profit-
ably can at present) the shape of the included surface.

We shall find it useful to restrict ourselves still
further, and only consider those boundaries which have
no rough points of the surface in them. Thus on the
surface of the cube we will only consider portions which
are entirely included in one of the plane faces ; on the
surface of the cylinder, only portions which are entirely
included in one of the flat faces, or in the curved part,
or which include one of the flat faces and part of the
curved portion.

This being so, the characteristics which we have to
remark in the boundaries of pieces of surface may be
sufficiently studied by means of figures drawn on paper.
We may bend the paper to assure ourselves that the
same general properties belong to figures on a cylinder,
and to make our ideas quite distinct it is worth while
to draw some on a sphere or other such surface.

In fig. 17 are some patches of surface ; a square, a
three-cornered piece, and two overlapping circles. For

Fig. 17.

distinctness, the part where the circles overlap is left
white, the rest being made black.

Attending now specially to the boundary of these
patches, we observe that it consists of smooth parts and
of corners or angles. Some of these corners project

and some are re-entrant. The pieces of surface are not
solid moveable things like the portions of space we
considered before, but we can in a measure imitate our
previous experiments by cutting out the figures with a
penknife, so as to leave their previous positions marked
by the holes. We shall then find, on applying the cut-
out pieces to one another, or to the holes, that at all
smooth-points the boundaries fit one another in a certain sense. Namely, if we place two smooth-points in
contact we cannot roll one figure on the other without
separating these points; whereas if we place a sharp-
point (or angle) on a smooth-point we can roll one figure
on the other without separating the points. If we
attempt to put two angles together without letting the
figures overlap, the same things may happen that we
found true in the case of the edges of solid bodies.
Suppose, for example, that we try to put an angle of the
square into one of the re-entrant angles of the figure
made by the two overlapping circles. If the re-entrant
angle is too sharp, we shall not be able to get it in at
all; this is the case of fig. 12. If it is wide enough,
the square will be able to rock in it; this is the case of
fig. 13. Between these two there is an intermediate
case in which one angle just fits the other; actual
contact takes place, and no rocking is possible. In
this case we say that the two angles are of the same
shape, or that they are *equal* to one another.

From all this we are led to conclude that *shape is a
matter of angles*, and that identity of shape depends on
equality of angle. We dealt with the size of a body by
considering a simple case of it, viz. length or distance,
and by measuring a sufficient number of lengths in different directions could find out all that is to be known
about the size of a body. It is, indeed, also true that a

F

knowledge of all the lengths which can be measured in a body would carry with it a knowledge of its shape; but still length is not in itself an element of shape. That which does the same for us in regard to shape that length does with regard to size, is angle. In other words, just as we say that two bodies are of the same size if to any line that can be drawn in the one there corresponds an exactly equal line in the other, so we say that two bodies are of the same shape, if to every angle that can be drawn on one of them there corresponds an exactly equal angle on the other.

Just as we measured lengths by a stick or a piece of tape so we measure angles with a pair of compasses; and two angles are said to be equal when they fit the same opening of the compasses. And as before, the statement that a thing can be moved about without altering its shape may be shown to amount only to this, that two angles which fit in one place will fit also in another, no matter how they have been brought from the one place to the other.

§ 5. *The Plane and the Straight Line.*

We have now to describe a particular kind of surface and a particular kind of line with which geometry is very much concerned. These are the *plane* surface and the *straight* line.

The plane surface may be defined as one which is of the same shape all over and on both sides. This property of it is illustrated by the method which is practically used to make such a surface. The method is to take three surfaces and grind them down until any two will fit one another all over. Suppose the three surfaces to be A, B, C; then, since A will fit B, it follows that the

space outside A is of the same shape as the space inside
B ; and because B will fit C, that the space inside B is of
the same shape as the space outside C. It follows therefore that the space outside A is of the same shape as the
space outside C. But since A will fit C when we put
them together, the space inside A is of the same shape
as the space outside C. But the space outside C was
shown to be of the same shape as the space outside A ;
consequently the space outside A is of the same shape as
the space inside ; and so, if three surfaces are ground
together so that each pair of them will fit, each of them
becomes a surface which is of the same shape on both
sides : that is to say, if we take a body which is partly
bounded by a plane surface, we can slide it all over this
surface and it will fit everywhere, and we may also turn
it round and apply it to the other side of the surface
and it will fit there too. This property is sometimes
more technically expressed by saying that a plane is a
surface which divides space into two *congruent regions.*

A straight line may be defined in a similar way. It
is a division between two parts of a plane, which two
parts are, so far as the dividing line is concerned, of the
same shape ; or we may say what comes to the same
effect, that a straight line is a line of the same shape all
along and on both sides.

A body may have two plane surfaces ; one part of it,
that is, may be bounded by one plane and another part
by another. If these two plane surfaces have a common
edge, this edge, which is called their *intersection,* is a
straight line. We may then, if we like, take as our
definition of a straight line that it is the intersection of
two planes.

It must be understood that when a part of the surface of a body is plane, this plane may be conceived as

extending beyond the body in all directions. For instance, the upper surface of a table is plane and horizontal. Now it is quite an intelligible question to ask about a point which is anywhere in the room whether it is higher or lower than the surface of the table. The points which are higher will be divided from those which are lower by an imaginary surface which is a continuation of the plane surface of the table. So then we are at liberty to speak of the line of intersection of two plane surfaces of a body whether these are adjacent portions of surface or not, and we may in every case suppose them to meet one another and to be prolonged across the edge in which they meet.

Leibniz, who was the first to give these definitions of a plane and of a straight line, gave also another definition of a straight line. If we fix two points of a body, it will not be entirely fixed, but it will be able to turn round. All points of it will then change their position excepting those which are in the straight line joining the two fixed points; and Leibniz accordingly defined a straight line as being the aggregate of those points of a body which are unmoved when it is turned about with two points fixed. If we suppose the body to have a plane face passing through the two fixed points, this definition will fall back on the former one which defines a straight line as the intersection of two planes.

It hardly needs any words to prove that the first two definitions of a plane are equivalent; that is, that two surfaces, each of which is of the same shape all over and on both sides, will have for their intersection a line which is of the same shape all along and on both sides. For if we slide each plane upon itself it will, being of the same shape all over, occupy as a whole the same unchanging position (*i.e.* wherever there was part of the planes before there will be part, though a different part, of the planes now), so that their line of intersection occupies the same position throughout (though the part of the line occupying any particular position is different). The line is therefore of the same shape all along. And in a similar way we can, without changing the position of the planes as a whole, move them so that the right-hand part of each shall become the left-hand part, and the upper part the lower; and this will amount to changing the line of intersection end for end. But this line is in the same place after the change as before; and it is therefore of the same shape on both sides.

From the first definition we see that two straight lines cannot coincide for a certain distance and then diverge from one another. For since the plane surface is of the same shape on the two sides of a straight line, we may take up the surface on one side and turn it over and it will fit the surface on the other side. If this is true of one of our supposed straight lines, it is quite clear that it cannot at the same time be true of the other; for we must either be bringing over more to fit less, or less to fit more.

§ 6. *Properties of Triangles.*

We can now reduce to a more precise form our first observation about space, that a body may be moved about in it without altering its size or shape. Let us suppose that our body has for one of its faces a *triangle*, that is to say, the portion of a plane bounded by three straight lines. We find that this triangle can be moved into any new position that we like, while the lengths of its sides and its angles remain the same ; or we may put the statement into the form that when any triangle is once drawn, another triangle of the same size and shape can be drawn in any part of space.

From this it will follow that if there are two triangles which have a side of the one equal to a side of the other, and the angles at the ends of that side in the one equal to the angles at the ends of the equal side in the other, then the two triangles are merely the same triangle in different positions; that is, they are of the same size and shape. For if we take the first triangle and so far put it into the position of the second that the two equal sides coincide, then because the angles at the ends of the one are respectively equal to those at the ends of the other, the remaining two sides of the first triangle will begin to coincide with the remaining two sides of the second. But we have seen that straight lines cannot begin to coincide and then diverge; and consequently these sides will coincide throughout and the triangles will entirely coincide.

Our second observation, that we may have things which are of the same shape but not of the same size, may also be made more precise by application to the case of triangles. It tells us that any triangle may be magnified or diminished to any degree without altering its angles, or that if a triangle be drawn, another triangle having the same angles may be drawn of any size in any part of space.

From this statement we are able to deduce two very important consequences. One is, that two straight lines cannot intersect in more points than one; and the other that, if two straight lines can be drawn in the same plane so as not to intersect at all, the angles they make with any third line in their plane which meets them, will be equal.

To prove the first of these, let A B and A C (fig. 18) be
two straight lines which meet at A. Draw a third line
B C, meeting both of them, and the three lines then form a
triangle. If we now make a point P travel along the line
A B it must, in virtue of our second observation, be always
possible to draw through this point a line which shall
meet A C in Q so as to make a triangle A P Q of the same

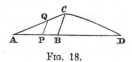

FIG. 18.

shape as A B C. But if the line A C were to meet A B in
some other point D besides A, then through this point
D it would clearly not be possible to draw a line so as
to make a triangle at all. It follows then that such
a point as D does not exist, and in fact that two
straight lines which have once met must go on diverg-
ing from each other and can never meet again.[1]

To prove the second, suppose that the lines A C and
B D (fig. 19) are in the same plane, and are such as

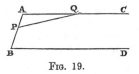

FIG. 19.

never to meet at all (in which case they are called
parallel), while the line A B meets them both. If we
make a point P travel along B A towards A, and, as it
moves, draw through it always a line making the same
angle with B A that B D makes with B A, then this

---

[1] This property might also be deduced from the first definition of a
straight line, by the method already used to show that two straight lines
cannot coincide for part of their length and then diverge.

moving line can never meet A C until it wholly coincides with it. For if it can, let P Q be such a position of the moving line; then it is possible to draw through B a line which, with A B and A C, shall form a triangle of the same shape as the triangle A P Q. But for this to be the case the line drawn through B must make the same angle with A B that P Q makes with it, that is, it must be the line B D. And the three lines B D, B A, A C cannot form a triangle, for B D and A C never meet. Consequently there can be no such triangle as A P Q, or the moveable line can never meet A C until it entirely coincides with it. But since this line always makes with B A the same angle that B D does, and in one position coincides with A C, it follows that A C makes with B A the same angle that B D does. This is the famous proposition about parallel lines.[1]

The first of these deductions will now show us that if two triangles have an angle of the one equal to an angle of the other and the sides containing these angles respectively equal, they must be equal in all particulars. For if we take up one of the triangles and put it down

---

[1] Two straight lines which cut one another form at the point where they cross four angles which are equal in pairs. It is often necessary to distinguish between the two different angles which the lines make with one another. This is done by the understanding that A B shall mean the line

(i)   (ii)

drawn from A to B, and B A the line drawn from B to A, so that the angle between A B and C D (i) is the angle B O D, but the angle between B A and C D (ii) is the angle D O A.

So the angle spoken of above as made by A C with B A is not the angle C A B (which is clearly, in general, unequal to the angle D B A), but the angle C A E, where E is a point in B A produced through A.

on the other so that these angles coincide and equal
sides are on the same side of them, then the containing sides will begin to coincide, and cannot therefore afterwards diverge. But as they are of the same
length in the one triangle as they are in the other, the
ends of them belonging to the one triangle will rest
upon the ends belonging to the other, so that the remaining sides of the two triangles will have their ends in
common and must therefore coincide altogether, since
otherwise two straight lines would meet in more points
than one. The one triangle will then exactly cover the
other ; that is to say, they are equal in all respects.

In the same way we may see that if two triangles
have two angles in the one equal to two angles in the
other, they are of the same shape. For one of them
can be magnified or diminished until the side joining
these two angles in it becomes of the same length as
the side joining the two corresponding angles in the
other ; and as no alteration is thereby made in the
shape of the triangle, it will be enough for us to prove
that the new triangle is of the same shape as the other
given triangle. But if we now compare these two, we
see that they have a pair of corresponding sides which
have been made equal, and the angles at the ends of
these sides equal also (for they were equal in the
original triangles, and have not been altered by the
change of size), so that we fall back on a case already
considered, in which it was shown that the third angles
are equal, and the triangles consequently of the same
shape.

If we apply these propositions not merely to two
different triangles but to the same triangle, we find
that if a triangle has two of its sides equal it will have
the two angles opposite to them also equal ; and that,

conversely, if it has two angles equal it will have the two sides opposite to them also equal; for in each of these cases the triangle may be turned over and made to fit itself. Such a triangle is called *isosceles.*

The theorem about parallel lines which we deduced from our second assumption about space leads very easily to a theorem of especial importance, viz. that the three angles of a triangle are together equal to two right angles.

If we draw through A, a corner of the triangle A B C (fig. 20), a line D A E, making with the side A C

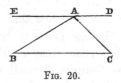

Fig. 20.

the same angle as B C makes with it, this line will, as we have proved, never meet B C, that is, it will be parallel to it. It will consequently make with A B the same angle as B C makes with it,[1] so that the three angles A B C, B A C, and B C A are respectively equal to the angles E A B, B A C, and C A D, and these three make up two right angles.

Another statement of this theorem is sometimes of use.

If the sides of a triangle be produced, what are called the *exterior angles* of the triangle are formed. If, for example, the side B C of the triangle A B C (fig. 21) is produced beyond C to D, A C D is an exterior angle of the triangle, while of the interior angles of the triangle A C B is said to be *adjacent,* and C A B and A B C to be *opposite* to this exterior angle. It is clear that as

---

[1] The convention mentioned in the last footnote must be remembered.

each side of the triangle may be produced in two directions, any triangle has six exterior angles.

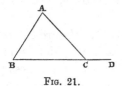

FIG. 21.

The other form into which our proposition may be thrown is that either of the exterior angles of a triangle is equal to the sum of the two interior angles opposite to it. For, in the figure, the exterior angle A C D, together with A C B, makes two right angles, and it must therefore be equal to the sum of the two angles which also make up two right angles with A C B.

§ 7. *Properties of Circles; Related Circles and Triangles.*

We may now apply this proposition to prove an important property of the circle, viz. that if we take two fixed points on the circumference of a circle and join them to a third point on the circle, the angle between the joining lines will depend only upon the first two points and not at all upon the third. If, for example, we join the points A, B (fig. 22) to C we shall show that, wherever on the circumference C may be, the angle A C B is always one-half of A O B; O being the centre of the circle.

Let C O produced meet the circumference in D. Then since the triangle O A C is isosceles, the angles O A C and O C A are equal, and so for a similar reason are the angles O B C and O C B.

But we have just shown that the exterior angle A O D is equal to the sum of the angles O A C and O C A;

and since these are equal to one another it must be double of either of them, say of o c a. Similarly the angle b o d is double of o c b, and consequently a o b is double of a c b.

In the case of the first figure (i) we have taken the sum of two angles each of which is double of another, and asserted that the sum of the first pair is twice the sum of the second pair; in the case of the second figure (ii) we have taken the difference of two angles

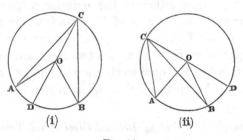

(i)   (ii)

Fig. 22.

each of which is double of another, and asserted that the difference of the first pair is twice the difference of the second pair.

Since therefore a c b is always half of a o b, wherever c may be placed in the upper of the two segments into which the circle is divided by the straight line a b, we see that the magnitude of this angle depends only on the positions of a and b, and not on the position of c.   But now let us consider what will happen if c is in the lower segment of the circle. As before, the triangles o a c and o b c (fig. 23) are isosceles, and the angles d o a and d o b are respectively double of o c a and o c b.   Consequently, the whole angle a o b formed by making o a turn round o into the position o b, so as to pass through the position o d (in the way, that is, in which the hands of a clock turn), this whole angle is double of A C B.

By our previous reasoning the angle A D B, formed by joining A and B to D, is one-half of the angle A O B, which is made by turning O B towards O A as the hands of a clock move. The sum of these two angles, each of which we have denoted by A O B, is a complete revolution about the point O ; in other words, is four

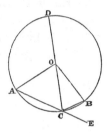

FIG. 23.

right angles. Hence the sum of the angles A D B, A C B, which are the halves of these, is two right angles. Or we may put the theorem otherwise, and say that the opposite angles of a four-sided figure whose angles lie on the circumference of a circle are together equal to two right angles.

We appear therefore to have arrived at two different statements according as the point C is in the one or the other of the segments into which the circle is divided by the straight line A B. But these statements are really the same, and it is easy to include them in one proposition. If we produce A C in the last figure to E, the angles A C B and B C E are together equal to two right angles ; and consequently B C E is equal to A D B. This angle B C E is the angle through which C B must be turned in the way the hands of a clock move,

so that its direction may coincide with that of A C. But
we may describe in precisely the same words the angle
A C B in fig. 22, where C was in the upper segment of the
circle; so that we may always put the theorem in these
words:—If A and B are fixed points on the circumference of a circle, and C any other point on it, the angle
through which C B must be turned clockwise in order to
coincide with C A or A C, whichever happens first, is
equal to half the angle through which O B must be
turned clockwise in order to coincide with O A.

We shall now make use of this to prove another interesting proposition. If three points D, E, F (fig. 24)

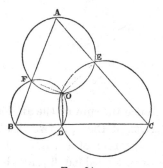

Fig. 24.

be taken on the sides of a triangle A B C, D being on B C,
E on C A, F on A B, then three circles can be drawn
passing respectively through A F E, B D F, C E D. These
three circles can be shown to meet in the same point O.
For let O in the first place stand for the intersection of
the two circles A F E and B F D, then the angles F A E
and F O E make up two right angles, and so do the
angles D O F and D B F. But the three angles at O make
four right angles, and the three angles of the triangle
A B C make two right angles; and of these six angles
two pairs have been shown to make up two right

angles each. Therefore the remaining pair, viz. the
angles D O E and D C E, make up two right angles. It
follows that the circle which goes through the points
C E D will pass through o, that is, the three circles all
meet in this point.

There is no restriction imposed on the positions of
the points D, E, F,[1] they may be taken either on the sides

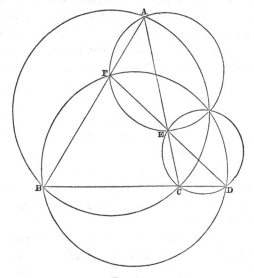

Fig. 25.

of the triangle or on those sides produced, and in particular we may take them to lie on any fourth straight
line D E F; and the theorem may be stated thus:—If
any four straight lines be taken (fig. 25), one of which
meets the triangle A B C formed by the other three in
the points D, E, F, then the circles through the points

[1] If either of the points D, E, F, is taken on a side produced, the proof
given above will not apply literally; but the necessary changes are slight
and obvious.

A F E, B D F, C E D meet in a point. But there is no reason why we should not take A F E as the triangle formed by three lines, and the fourth line D C B as the line which cuts the sides of this triangle. The proposition is equally true in this case, and it follows that the circles through A B C, E C D, F B D will meet in one point. This must be the same point as before, since two of the circles of this set are the same as two of the previous set; consequently all four circles meet in a point, and we can now state our proposition as follows :

Given four straight lines, there can be formed from them four triangles by leaving out each in turn; the circles which circumscribe these four triangles meet in a point.

This proposition is the third of a series.

If we take any two straight lines they determine a point, viz. their point of intersection.

If we take three straight lines we get three such points of intersection; and these three determine a circle, viz. the circle circumscribing the triangle formed by the three lines.

Four straight lines determine four sets of three lines by leaving out each in turn; and the four circles belonging to these sets of three meet in a point.

In the same way five lines determine five sets of four, and each of these sets of four gives rise, by the proposition just proved, to a point. It has been shown by Miquel, that these five points lie on the same circle.

And this series of theorems has been shown [1] to be endless. Six straight lines determine six sets of five by leaving them out one by one. Each set of five has, by

---

[1] By Prof. Clifford himself in the *Oxford, Cambridge, and Dublin Messenger of Mathematics*, vol. v. p. 124. See his *Mathematical Papers*, pp. 51–54.

Miquel's theorem, a circle belonging to it. These six circles meet in the same point, and so on for ever. Any even number (2n) of straight lines determines a point as the intersection of the same number of circles. It we take one line more, this odd number $(2n+1)$ determines as many sets of $2n$ lines, and to each of these sets belongs a point; these $2n+1$ points lie on a circle.

§ 8. *The Conic Sections.*

The shadow of a circle cast on a flat surface by a luminous point may have three different shapes. These are three curves of great historic interest, and of the utmost importance in geometry and its applications. The lines we have so far treated, viz. the straight line and circle, are special cases of these curves; and we may naturally at this point investigate a few of the properties of the more general forms.

If a circular disc be held in any position so that it is altogether below the flame of a candle, and its shadow be allowed to fall on the table, this shadow will be of an oval form, except in two extreme cases, in one of which it also is a circle, and in the other is a straight line. The former of these cases happens when the disc is held parallel to the table, and the latter when the disc is held edgewise to the candle; or, in other words, is so placed that the plane in which it lies passes through the luminous point. The oval form which, with these two exceptions, the shadow presents is called an *ellipse* (i). The paths pursued by the planets round the sun are of this form.

If the circular disc be now held so that its highest point is just on a level with the flame of the candle, the shadow will as before be oval at the end near the candle;

G

but instead of closing up into another oval end as we move away from the candle, the two sides of it will continue to open out without any limit, tending however to become more and more parallel. This form of the shadow is called a *parabola* (ii). It is very nearly the orbit of many comets, and is also nearly represented by the path of a stone thrown up obliquely. If there were no atmosphere to retard the motion of the stone it would exactly describe a parabola.

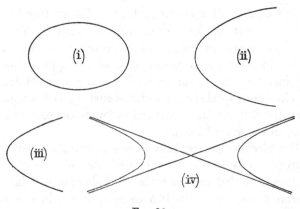

FIG. 26.

If we now hold the circular disc higher up still, so that a horizontal plane at the level of the candle flame divides it into two parts, only one of these parts will cast any shadow at all, and that will be a curve such as is shown in the figure, the two sides of which diverge in quite different directions, and do not, as in the case of the parabola, tend to become parallel (iii).

But although for physical purposes this curve is the whole of the shadow, yet for geometrical purposes it is not the whole. We may suppose that instead of being a shadow our curve was formed by joining the luminous

point by straight lines to points round the edge of the
disc, and producing these straight lines until they meet
the table.

This geometrical mode of construction will equally
apply to the part of the circle which is above the candle
flame, although that does not cast any shadow. If we
join these points of the circle to the candle flame, and
prolong the joining lines beyond it, they will meet the
table on the other side of the candle, and will trace out
a curve there which is exactly similar and equal to the
physical shadow (iv). We may call this the *anti-shadow*
or *geometrical shadow* of the circle. It is found that for
geometrical purposes these two branches must be considered as forming only one curve, which is called an
hyperbola. There are two straight lines to which the
curve gets nearer and nearer the further away it goes
from their point of intersection, but which it never
actually meets. For this reason they are called *asymptotes*, from a Greek word meaning 'not falling together.' These lines are parallel to the two straight
lines which join the candle flame to the two points of
the circle which are level with it.

We saw some time ago that a surface was formed
by the motion of a line. Now if a right line in its
motion always passes through one fixed point, the surface
which it traces out is called a *cone*, and the fixed point
is called its *vertex*. And thus the three curves which we
have just described are called *conic sections*, for they
may be made by cutting a cone by a plane. In fact, it
is in this way that the shadow of the circle is formed;
for if we consider the straight lines which join the
candle flame to all parts of the edge of the circle we see
that they form a cone whose vertex is the candle flame
and whose base is the circle.

We must suppose these lines not to end at the flame but to be prolonged through it, and we shall so get what would commonly be called two cones with their points together, but what in geometry is called one *conical surface* having two *sheets*. The section of this conical surface by the horizontal plane of the table is the shadow of the circle; the sheet in which the circle lies gives us the ordinary physical shadow, the other sheet (if the plane of section meets it) gives what we have called the geometrical shadow.

The consideration of the shadows of curves is a method much used for finding out their properties, for there are certain geometrical properties which are always common to a figure and its shadow. For example, if we draw on a sheet of glass two curves which cut one another, then the shadows of the two curves cast through the sheet of glass on the table will also cut one another. The shadow of a straight line is always a straight line, for all the rays of light from the flame through various points of a straight line lie in a plane, and this plane meets the plane surface of the table in a straight line which is the shadow. Consequently if any curve is cut by a straight line in a certain number of points, the shadow of the curve will be cut by the shadow of the straight line in the same number of points. Since a circle is cut by a straight line in two points or in none at all, it follows that any shadow of a circle must be cut by a straight line in two points or in none at all.

When a straight line touches a circle the two points of intersection coalesce into one point. We see then that this must also be the case with any shadow of the circle. Again, from a point outside the circle it is possible to draw two lines which touch the circle; so from a point outside either of the three curves which we have just described, it is possible to draw two lines to touch the curve. From a point inside the circle no tangent can be drawn to it, and accordingly no tangent can be drawn to any conic section from a point inside it.

This method of deriving the properties of one curve from those of another of which it is the shadow, is called the method of *projection*.

The particular case of it which is of the greatest use is that in which we suppose the luminous point by which the shadow is cast to be ever so far away. Suppose, for example, that the shadow of a circle held obliquely is cast on the table by a star situated directly overhead, and at an indefinitely great distance. The lines joining the star to all the points of the circle will then be vertical lines, and they will no longer form a cone but a cylinder. One of the chief advantages of this kind of projection is that the shadows of two parallel lines will remain parallel, which is not generally the case in the other kind of projection. The shadow of the circle which we obtain now is always an ellipse; and we are able to find out in this way some very important properties of the curve, the corresponding properties of the circle being for the most part evident at a glance on account of the symmetry of the figure.

For instance, let us suppose that the circle whose shadow we are examining is vertical, and let us take a vertical diameter of it, so that the tangents at its ends are horizontal. It will be clear from the symmetry of the figure that all horizontal lines in it are divided into two equal parts by the vertical diameter, or we may say that the diameter of the circle bisects all chords parallel to the tangents at its extremities. When the shadow of this figure is cast by an infinitely distant star (which we must not now suppose to be directly overhead, for then the shadow would be merely a straight line), the point of bisection of the shadow of any straight line is the shadow of the middle point of that line, and thus we learn that it is true of the ellipse that any line which joins the points of contact of parallel tangents bisects all chords parallel to those tangents. Such a line is, as in the case of the circle, termed a *diameter*. Since the shadow of a diameter of the circle is a diameter of the ellipse, it follows that all diameters of the ellipse pass through one and the same point, namely, the shadow of the centre of the circle; this common intersection of diameters is termed the *centre* also of the ellipse.

Again, a horizontal diameter in the circle just considered will bisect all vertical chords, and thus we see that if one diameter bisects all chords parallel to a second, the second will bisect all chords parallel to the first.

The method of projection tells us that this is also true of the ellipse. Such diameters are called *conjugate diameters*, but they are no longer at right angles in the ellipse as they were in the case of the circle.

Since the shadow of a circle which is cast in this way by an infinitely distant point is always an ellipse, we cannot use the same method in order to obtain the properties of the hyperbola. But it is found by other methods that these same statements are true of the hyperbola which we have just seen to be true of the ellipse. There is however this great difference between the two curves. The centre of the ellipse is inside it, but the centre of the hyperbola is outside it. Also all lines drawn through the centre of the ellipse meet the curve in two points, but it is only certain lines through the centre of the hyperbola which meet the curve at all. Of any two conjugate diameters of the hyperbola one meets the curve and the other does not. But it still remains true that each of them bisects all chords parallel to the other.

§ 9. *On Surfaces of the Second Order.*

We began with the consideration of the simplest kind of line and the simplest kind of surface, the straight line and the plane; and we have since found out some of the properties of four different curved lines —the circle, the ellipse, the parabola, and the hyperbola. Let us now consider some curved surfaces; and first, the surface analogous to the circle. This surface is the *sphere*. It is defined, as a circle is, by the property that all its points are at the same distance from the centre.

Perhaps the most important question to be asked about a surface is, What are the shapes of the curved lines in which it is met by other surfaces, especially in the case when these other surfaces are planes? Now a plane which cuts a sphere cuts it, as can easily be shown, in a circle. This circle, as we move the plane further and further away from the centre of the sphere, will get smaller and smaller, and will finally contract into a point. In this case the plane is said to *touch* the sphere; and we notice a very obvious but important fact, that the sphere then lies entirely on one side of the plane. If the plane be moved still further away from the centre it will not meet the sphere at all.

Again, if we take a point outside the sphere we can draw a number of planes to pass through it and touch the sphere, and all the points in which they touch it lie on

a circle. Also a cone can be drawn whose vertex is
the point, and which touches the sphere all round the
circle in which these planes touch it. This is called
the *tangent-cone* of the point. It is clear that from a
point inside the sphere no tangent-cone can be drawn.

Similar properties belong also to certain other surfaces which resemble the sphere in the fact that they
are met by a straight line in *two* points at most; such
surfaces are on this account called of the *second order*.

Just as we may suppose an ellipse to be got from
a circle by pulling it out in one direction, so we may
get a *spheroid* from a sphere either by pulling it out so
as to make a thing like an egg, or by squeezing it so
as to make a thing like an orange. Each of these
forms is symmetrical about one diameter, but not about
all. A figure like an orange, for example, or like the
earth, has a diameter through its poles less than any
diameter in the plane of its equator, but all diameters in
its equator are equal. Again, a spheroid like an egg
has all the diameters through its equator equal to one
another, but the diameter through its poles is longer
than any other diameter.

If we now take an orange or an egg and make its
equator into an ellipse instead of a circle, say by pulling out the equator of the orange or squeezing the
equator of the egg, so that the surface has now three
diameters at right angles all unequal to one another,
we obtain what is called an *ellipsoid*. This surface
plays the same part in the geometry of surfaces that the
ellipse does in the geometry of curves. Just as every
plane which cuts a sphere cuts it in a circle, so every
plane which cuts an ellipsoid cuts it in an ellipse. It
is indeed possible to cut an ellipsoid by a plane so that
the section shall be a circle, but this must be regarded

as a particular kind of ellipse, viz. an ellipse with
two equal axes. Again, just as was the case with the
sphere, we can draw a set of planes through an external point all of which touch the ellipsoid. Their points
of contact lie on a certain ellipse, and a cone can be
drawn which has the external point for its vertex and
touches the ellipsoid all round this ellipse. The ellipsoid resembles a sphere in this respect also, that when

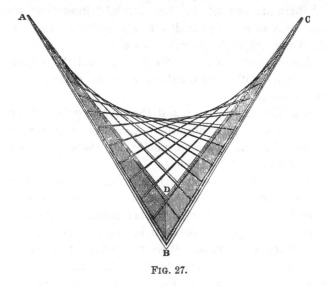

FIG. 27.

it is touched by a plane it lies wholly on one side of
that plane.

There are also surfaces which bear to the hyperbola
and the parabola relations somewhat similar to those
borne to the circle by the sphere, and to the ellipse by
the ellipsoid. We will now consider one of them, a
surface with many singular properties.

Let A B C D be a figure of card-board having four
equal sides, and let it be half cut through all along B D,

so that the triangles A B D, C B D can turn about the line
B D. Then let holes be made along the four sides of it at
equal distances, and let these holes be joined by threads
of silk parallel to the sides. If now the figure be bent
about the line B D and the silks are pulled tight it will
present an appearance like that in fig. 27, resembling
a saddle, or the top of a mountain pass.

This surface is composed entirely of straight lines,
and there are two sets of these straight lines; one set
which was originally parallel to A B, and the other set
which was originally parallel to A D.

A section of the figure through A C and the middle
point of B D will be a parabola with its concave side
turned upwards.

A section through B D and the middle point of A C
will be another parabola with its concave side turned
downwards, the common vertex of these parabolas
being the summit of the pass.

The tangent plane at this point will cut the surface
in two straight lines, while part of the surface will be
above the tangent plane and part below it. We may
regard this tangent plane as a horizontal plane at the
top of a mountain pass. If we travel over the pass, we
come up on one side to the level of the plane and then
go down on the other. But if we go down from a
mountain on the right and go up the mountain on the
left, we shall always be above the horizontal plane. A
section by a horizontal plane a little above this tangent
plane will be a hyperbola whose asymptotes will be
parallel to the straight lines in which the tangent plane
meets the surface. A section by a horizontal plane a
little below will also be a hyperbola with its asymptotes
parallel to these lines, but it will be situated in the
other pair of angles formed by these asymptotes. If

we suppose the cutting plane to move downwards from a position above the tangent plane (remaining always horizontal), then we shall see the two branches of the first hyperbola approach one another and get sharper and sharper until they meet and become simply two crossing straight lines. These lines will then have their corners rounded off and will be divided in the other direction and open out into the second hyperbola.

This leads us to suppose that a pair of intersecting straight lines is only a particular case of a hyperbola, and that we may consider the hyperbola as derived from the two crossing straight lines by dividing them at their point of intersection and rounding off the corners.

§ 10. *How to form Curves of the Third and Higher Orders.*

The method of the preceding paragraph may be extended so as to discover the forms of new curves by putting known curves together. By a mode of expression which sounds paradoxical, yet is found convenient, a straight line is called a curve of the first order, because it can be met by another straight line in only one point; but two straight lines taken together are called a curve of the second order, because they can be met by a straight line in two points. The circle, and its shadows, the ellipse, parabola, and hyperbola, are also called curves of the second order, because they can be met by a straight line in two points, but not in more than two points; and we see that by this process of rounding off the corners and the method of projection we can derive all these curves of the second order from a pair of straight lines.

A similar process enables us to draw curves of the third order. An ellipse and a straight line taken together form a curve of the third order. If now we round off the corners at both the points where they meet we obtain (fig. 28) a curve consisting of an oval and a sinuous portion called a ' snake.' Now just as when we move a plane which cuts a sphere away from the centre, the curve of intersection shrinks up into a

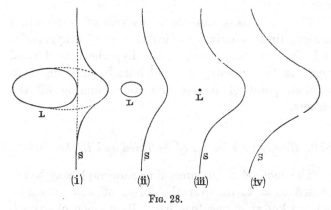

FIG. 28.

(i.) Full loop and snake.   (iii.) The loop has shrunk to a point.
(ii.) Shrunk loop and snake.   (iv.) Snake only.

point and then disappears, so we can vary our curve of the third order so as to make the oval which belongs to it shrink up into a point, and then disappear altogether, leaving only the sinuous part, but no variation will get rid of the ' snake.'

We may, if we like, only round off the corners at one of the intersections of the straight line and the ellipse, and we then have a curve of the third order crossing itself, having a *knot* or *double point* (fig. 29) ; and we can further suppose this loop to shrink up, and the curve will then be found to have a sharp point or *cusp*.

It was shown by Newton that all curves of the third order might be derived as shadows from the five forms

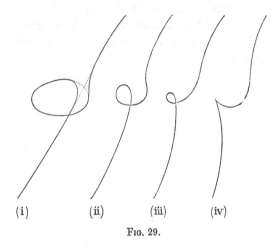

(i)　　　　(ii)　　　(iii)　　　(iv)

Fɪɢ. 29.

which we have just mentioned, viz. the oval and snake, the point and snake, the snake alone, the form with a knot, and the form with a cusp.

In the same way curves of the fourth order may be got by combining together two ellipses. If we suppose

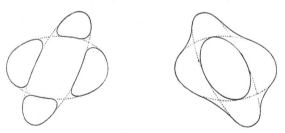

Fɪɢ. 30.

them to cross each other in four points we may round off all the corners at once and so obtain two different forms, either four ovals all outside one another or an oval with four dints in it, and another oval inside it (fig. 30).

But the number of forms of curves of the fourth order is so great that it has never yet been completely catalogued; and curves of higher orders are of still more varied shapes.

CHAPTER III.

QUANTITY.

§ 1. *The Measurement of Quantities.*

WE considered at the beginning of the first chapter,
on Number, the process of counting things which are
separate from one another, such as letters or men or
sheep, and we found it to be a fundamental property of
this counting that the result was not affected by the
order in which the things to be counted were taken;
that one of the things, that is, was as good as another
at any stage of the process.

We may also count things which are not separate
but all in one piece. For example, we may say that a
room is sixteen feet broad. And in order to count the
number of feet in the breadth of this room we should
probably take a foot rule and measure off first a foot
close to the wall, then another beginning where that
ended, and so on until we reached the opposite wall.
Now when these feet are thus marked off they may,
just like any other separate things, be counted in
whatever order we please, and the number of them
will always be sixteen.

But this is not all the variety in the process of
counting which is possible. For suppose that we take
a stick whose length is equal to the breadth of the
room. Then we may cut out a foot of it wherever we
please, and join the ends together. And if we then

cut out another foot from any part of the remainder
and join the ends, and repeat the process fifteen times,
we shall find that there will always be a foot length
left when the last two ends are joined together. So,
when we are counting things that are all in one piece,
like the length of the stick or the breadth of the room,
not only is the order in which we count the feet immaterial, but also the position of the actual feet which
we count.

Again, if we say that a packet contains a pound, or
sixteen ounces, of tea, we mean that if we take any
ounce of it out, then any other ounce out of what is
left, and so on until we have taken away fifteen ounces,
there will always be an ounce left.

If I say that I have been writing for fifteen minutes
it will of course have been impossible actually to count
these minutes except in the order in which they really
followed one another, but it will still be true that, if
any separate fourteen minutes had been marked off
during that interval of time, the remainder of it, made
up of the interstices between these minutes, would
amount on the whole to one minute.

In all these cases we have been counting things that
hang together in one piece; and we find that we may
choose at will not only the order of counting but even
the things that we count without altering the result.
This process is called the *measurement of quantities*.

But now suppose that when we measure the breadth
of a room we find it to be not sixteen feet exactly, but
sixteen feet and something over. It may be sixteen
feet and five inches. And if so, in order to measure
the something over, we merely repeat the same process as before; only that instead of counting feet we
count inches, which are smaller than feet. If the

breadth is found not to be an exact number of inches, but that something is left beside the five inches, we might measure that in eighths of an inch. There might, for example, be three eighths of an inch over. But there is no security that the process will end here; for the breadth of the room may not contain an exact number of eighths of an inch. Still it may be said that nobody wants to know the breadth of a room more exactly than to within an eighth of an inch.

Again, when we measure a quantity of tea it may be nearly, but not exactly, sixteen ounces; there may be something over. This remainder we shall then measure in grains. And here, as before, we are repeating the same process by which we count things which are all in one piece; only we count grains, which are smaller things than ounces. There may still not be an exact number of grains in the packet of tea, but then nobody wants to know the weight of a packet of tea so nearly as to a grain.

And it is the same with time. A geological period may, if we are very accurate, be specified in hundreds of centuries; the length of a war in years; the time of departure of a train to within a minute; the moment of an eclipse to a second; our care being, in each case, merely to secure that the measurement is accurate enough for the purpose we have in hand.

To sum up. There is in common use a rough or approximate way of describing quantities, which consists in saying how many times the quantity to be described contains a certain standard quantity, and in neglecting whatever may remain. The smaller the standard quantity is the more accurate is the process, but it is in general no better than an approximation.

If then we want to describe a quantity accurately

H

and not by a mere approximation, what are we to do?
There is no way of doing this in words; the only possible method is to carry about either the quantity itself
or some other quantity which shall serve to represent
it. For instance, to represent the exact length and
breadth of a room we may draw it upon a scale of, say,
one inch to a foot and carry this drawing about.

Here we are representing a length by means of
another length; but it is not necessary to represent
weights by means of weights, or times by means of
times; they are both in practice represented by lengths.
When a chemist, wishing to weigh with great delicacy,
has gone as near as he can with the drachms which he
puts into his scales, he hangs a little rider upon the
beam of the scale, and the distance of this rider from
the middle indicates how much weight there is over.
And, if we suppose the balance to be perfectly true,
and that no friction or other source of error has to be
taken into account, it indicates this weight with real
accuracy.

Here then is a case in which a weight is indicated
by a length, namely, the distance from the centre of the
scale to the rider. Again, we habitually represent time
by means of a clock, and in this case the minute hand
moves by a succession of small jerks, possibly twice a
second. Such a clock will only reckon time in half
seconds, and can tell us nothing about smaller intervals
than this. But we may easily conceive of a clock in
which the motion of the minute hand is steady, and not
made by jerks. In this case the interval of time since
the end of the last hour will be accurately represented
by the length round the outer circle of the clock
measured from the top of it to the point of the minute
hand. And we notice that here also the quantity

which is measured in this way by a length is probably not the whole quantity which was to be estimated, but only that which remains over after the greater part has been counted by reference to some standard quantity.

We may thus describe weight and time, and indeed quantities of any kind whatever, by means of the lengths of lines; and in what follows, therefore, we shall only speak of quantities of length as completely representing measurable things of any sort.

§ 2. *The Addition and Subtraction of Quantities.*

For the addition of two lengths it is plainly sufficient to place them end to end in the same line. And we must notice that, as was the case with counting, so now, the possible variety in the mode of adding is far greater in the case of two quantities than in the case of two numbers. For either of the lengths, the aggregate of which we wish to measure, may be cut up into any number of parts, and these may be inserted at any points we please of the other length, without any change in the result of our addition.

Or the same may be seen, perhaps more clearly, by reference to the idea of ' steps.' Suppose we have a straight line with a mark upon it agreed on as a starting-point, and a series of marks ranged at equal distances along the line and numbered 1, 2, 3, 4. . . . Then any particular number is shown by making an index point to the right place on the line. And to add or subtract any other number from this, we have only to make the index move forwards or backwards over the corresponding number of divisions. But in the case of lengths we are not restricted to the places which are marked on the scale. Any length is shown by carrying the index to a place whose distance from the starting-point is the length in question (of which places there may be as many as we please between any two points which correspond to consecutive numbers), and another length is added or subtracted by making the index take a 'step' forwards or backwards of the necessary amount.

It is seen at once that, for quantities in general as well as for numbers, a succession of given steps may be made in any order we please and the result will always be the same.

§ 3. *The Multiplication and Division of Quantities.*

We have already considered cases in which a quantity is *multiplied*; that is to say, in which a certain number of equal quantities are added together, a process called the *multiplication* of one of them by that number. Thus the length sixteen feet is the result of multiplying one foot by sixteen.

We may now ask the inverse question: Given two lengths, what number must be used to multiply one of them in order to produce the other? And it has been implied in what we have said about the measurement of quantities that it is only in special cases that we can find a number which will be the answer to this question. If we ask, for example, by what number a foot must be multiplied in order to produce fifteen inches, the word 'number' requires to have its meaning altered and extended before we can give an answer. We know that an inch must be multiplied by fifteen in order to become fifteen inches. We may therefore first ask by what a foot must be multiplied in order to produce an inch. And the question seems at first absurd; because an inch must be multiplied by twelve in order to give a foot, and a foot has to be, not multiplied at all, but divided by twelve, in order to become an inch.

In order then to turn a foot into fifteen inches, we must go through the following process; we must divide it into twelve equal parts and take fifteen of them ; or, shortly, divide by twelve and multiply by fifteen. Or we may produce the same result by performing the steps of our process in the other order : we may first multiply by fifteen, so that we get fifteen feet, and then divide this length into twelve equal parts, each of which will be fifteen inches.

Now if instead of inventing a new name for this compound operation we choose to call it by the old name of multiplication, we shall be able to speak of multiplying a foot so as to get fifteen inches. The operation of multiplying by fifteen and dividing by twelve is written thus: $\frac{15}{12}$; and· so, to change a foot into fifteen inches, we multiply by the *fraction* $\frac{15}{12}$. Of this fraction the upper number (15) is termed the *numerator*, the lower (12) the *denominator*.

Now it was explained in the first chapter, that the formulæ of arithmetic and algebra are capable of a double interpretation. For instance, such a symbol as 3 meant, in the first place, a number of letters or men, or any other things ; but afterwards was regarded as meaning an operation, namely, that of trebling anything. And so now the symbol $\frac{15}{12}$ may be taken either as meaning ' so much' of a foot, or as meaning the operation by which a foot is changed into fifteen inches.

The degree in which one quantity is greater or less than another ; or, to put it more precisely, that amount of stretching or squeezing which must be applied to the latter in order to produce the former, is called the *ratio* of the two quantities. If a and b are any two lengths, the ratio of a to b is the operation of stretching or squeezing which will make b into a; and this operation can be always approximately, and sometimes exactly, represented by means of numbers.

§ 4. *The Arithmetical Expression of Ratios.*

For the approximate expression of ratios there are two methods in use. In each, as in measuring quantities in general, we proceed by using standards which are taken smaller and smaller as we go on. In the first, these standards are chosen according to a fixed law; in the second, our choice is suggested by the particular ratio which we are engaged in measuring.

The first method consists in using a series of standards each of which is a tenth part of the preceding. Thus to express the ratio of fifteen inches to a foot, we proceed thus. The fifteen inches contain a foot once, and there is a piece of length three inches, or a quarter of a foot, left over. This quarter of a foot is then measured in tenths of a foot, and we find that it is 2-tenths, with a piece—which proves to be half a tenth —over. So, if we chose to neglect this half-tenth we should call the ratio 12-tenths, or as we write it 1·2. But if we do not neglect the half-tenth, it has to be measured in hundredths of a foot; of which it makes 5 exactly. So that the result is 125 hundredths, or 1·25, accurately.

Again we will try to express in this way the length of the diagonal of a square in terms of a side. We find at once that the diagonal contains the side once, with a piece over: so that the ratio in question is 1 together with some fraction. If we now measure this remaining piece in tenth parts of a side we shall find that it contains

4 of them, with something left. Thus the ratio of the diagonal to the side may be approximately expressed by 14-tenths, or 1·4. If we now measure the piece left over in hundredth parts of the side we shall find that it contains one and a bit. Thus 141-hundredths, or 1·41 is a more accurate description of the ratio. And this bit can be shown to contain 4-thousandths of the side, and a bit over; so that we arrive at a still more accurate value, 1414-thousandths, or 1·414. And this process might be carried on to any degree of accuracy that was required; but in the present case, unlike that considered before, it would never end; for the ratio of the diagonal of a square to its side is one which cannot be accurately expressed by means of numbers.

The other method of approximation differs from the one just explained in this respect—that the successively smaller and smaller standard quantities in terms of which we measure the successive remainders are not fixed quantities, an inch, a tenth of an inch, a hundredth of an inch, and so on; but are suggested to us in the course of the approximation itself.

We begin, as we did before, by finding how many times the lesser quantity is contained in the greater, say, the side of a square in its diagonal. The answer in this case is, once and a piece over. Let the piece left over be called a. We then go on to try how many times this remainder, a, is contained in the side of the square. It is contained twice, and there is a remainder, say b. We then find how many times b is contained in a. Again twice, with a piece over, say c. And this process is repeated as often as we please, or until no remainder is left. It will, in the present case, be found that each remainder is contained twice, with something over, in the previous remainder.

Let us now inquire how this process enables us to find successive approximations to the ratio of the diagonal to the side of the square.

Suppose, first, that the piece a had been exactly half the length of the side; that is, that we may neglect the remainder b. Then the diagonal would be equal to the side together with half the side, that is, to three-halves of the side.

Next let us include b in our approximation, but neglect c; that is, let us suppose that b is exactly one half of a. Then the side contains a twice, and half of a; that is to say, contains five-halves of a; or a is two-fifths of the side. But the diagonal contains the side together with a, that is, contains the side and two-fifths of the side, or seven-fifths of the side. The piece neglected is here less than b, and b is one-fifth of the side of the square.

Again, let us include c in our approximation, and suppose it to be exactly one half of b. Then a, which contains b twice with c over, will be five-halves of b, that is b will be two-fifths of a. Hence the side will contain twice a and two-fifths of a, that is, twelve-fifths of a; so that a is five-twelfths of the side. And the diagonal is equal to the side together with a; that is, to seventeen-twelfths of the side. Also this approximation is closer than the preceding, for the piece neglected is now less than c, which is one-half of b, which is two-fifths of a, which is five-twelfths of the side; so that it is less than one-twelfth of the side.

By continuing this process we may find an approximation of any required degree of accuracy.

The first method of approximation is called the method of *decimals*; the second, that of *continued fractions*.

§ 5. *The Fourth Proportional.*

One of the chief differences between quantities and
numbers is that, while the division of one number by
another is only possible when the first number happens
to be a multiple of the other, in the case of quantities
it appears, and we are indeed accustomed to assume,
that any quantity may be divided by any number we
like; that is to say, any length—quantities of all kinds
being represented by lengths—may be divided into any
given number of equal parts. And, if division is always
possible, that compound operation made up of multiplication and division which we have called ' multiplying by a fraction' must also be always possible; for
example, we can find five-twelfths not only of a foot
but of any other length that we like.

The question now naturally arises whether that
general operation of stretching or squeezing which we
have called a *ratio* can be applied to all quantities alike.
If we have three lengths, *a*, *b*, *c*, there is a certain
operation of stretching or squeezing which will convert
a into *b*. Can the same operation be performed upon *c*
with the result of producing a fourth quantity *d*, such
that the ratio of *c* to *d* shall be the same as the ratio of
a to *b*? We assume that this quantity—the *fourth
proportional*, as it is called—does always exist; and
this assumption, as it really lies at the base of all
subsequent mathematics, is of so great importance as
to deserve further study.

We shall find that it is really included in the second
of the two assumptions that we made in the chapter
about space; namely, that figures of the same shape
may be constructed of different sizes. We found, in

considering this point, that it was sufficient to take the
case of triangles of different sizes of which the angles
were equal; and showed that one triangle might be
made into another of the same shape by the equal
magnifying of all its three sides; that is to say, when
two triangles have the same angles, the three ratios of
either side of one to the corresponding side of the other
are equal. If this is true, it is clear that the problem of
finding the fourth proportional is reduced to that of drawing two triangles of the same shape. Thus, for example,
let A B and A C represent the first two given quantities,
and A D the third (fig. 31); and let it be required to
find that quantity which is got from A D by the same

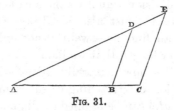

FIG. 31.

operation of stretching as is required to turn A B into
A C. Suppose that we join B D, and draw the line C E
making the angle A C E equal to the angle A B D. The two
triangles A B D and A C E are now of the same shape, and
consequently A C E can be got from A B D by the equal
stretching of all its sides; that is to say, the stretching
which makes A B into A C is the same as the stretching
which makes A D into A E. A E is therefore the fourth proportional required.

To render these matters clearer, it is well that we
should get a more exact notion of what we mean by the
fourth proportional. We have so far only described it
as something which is got from A D by the same process
which makes A B into A C. In what way are we to tell
whether the process is the same? We might, if we

liked, give a geometrical definition of it, founded upon
the construction just explained; and say that the ratio
of A D to A E shall be called ' equal ' to the ratio of A B to
A C, when triangles of the same shape can have for their
respective sides the lengths A B, A D, A C, and A E. But it
is better, if we can do it, to keep the science of quantity
distinct from the science of space, and to find some
definition of the fourth proportional which depends
upon quantity alone. Such a definition has been found,
and it is very important to notice the nature of it. For
we shall find that similar definitions have to be given of
other quantities whose existence is assumed by what is
called *the principle of continuity*. This principle is
simply the assumption, which we have stated already,
that all quantities can be divided into any given number
of equal parts.

If we apply two different operations of stretching
to the same quantity, that which produces the greater
result is naturally looked upon as an operation which
under like circumstances will always produce a greater
effect. Now we will make our definition of the fourth
proportional depend upon the very natural assumption
that, if two processes of stretching are applied to two
different quantities, that process which produces the
greater result in the one case will also produce the
greater result in the other.

Suppose now that we have tried to approximate to
the ratio which A C bears to A B, and that we have
found that A C is between seventeen-twelfths and
eighteen-twelfths of A B, then we have two processes
of stretching which can be applied to A B, the process
denoted by $\frac{17}{12}$ (that is, multiplying by 17 and dividing
by 12), and the process which makes A C of it. The
result of the former process is, by hypothesis, less than

the result of the latter, because A C is more than seventeen-twelfths of A B. Let us now apply these two processes to A D. The former will produce seventeen-twelfths of A D, the latter will produce the fourth proportional required. Consequently this fourth proportional must be greater than seventeen-twelfths of A D.

But we know further that A C is less than eighteen-twelfths of A B. Then the operation which makes A B into A C gives a less result than the operation of multiplying by 18 and dividing by 12. Let us now perform both upon A D. It will follow that the fourth proportional required is less than eighteen-twelfths of A D. The same thing will be true of any fractions we like to take, and we may state our result in this general form :—

According as A C is greater or is less than any specified fraction of A B, so will the fourth proportional (if it exists) be greater or be less than the same fraction of A D.

But we shall now show that this property is of itself sufficient to define, without ambiguity, the fourth

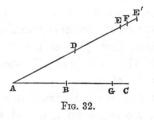

Fig. 32.

proportional; that is to say, we shall show that there cannot be two different lengths satisfying this condition at the same time.

If possible, let there be two lengths, A E and A E′, each of them a fourth proportional to A B, A C, A D (fig. 32). Then by taking a sufficient number of lengths each

equal to E E′, the sum of them can be made greater
than A D. Suppose for example that 500 of them
just fell short of the length A D, and that 501 exceeded
it; then, if we divide A D into 501 equal parts, each of
these parts will be less than E E′. Secondly, if we go
on marking off lengths from D towards E, each equal
to one of these small parts of A D, one of the points of
division must fall between E and E′; since E E′ is
greater than the distance between two of them. Let
this point of division be at F. Then A F is got from
A D by multiplying by some number or other and then
dividing by 501. If we apply this same process to A B
we shall arrive at a length A G, which must be either
greater or less than A C. If it is less than A C, then the
operation by which the length A B is made into A G is a
less amount of stretching than the operation by which
A B is made into A C. Consequently the operation
which turns A D into A F is a less amount of stretching
than that which gets A E, and also less than that which
gets A E′ from A D. Therefore A F must be less than A E,
and also less than A E′. But this is impossible, because
F lies between E and E′. And the argument would be
similar if we had supposed A G greater than A C.

Thus we have proved that there is only one length
that satisfies the condition that the process of making
A D into it is greater than all the fractions which are
less than the process of making A B into A C, and less
than all the fractions which are greater than this same
process.

Let us note more carefully the nature of this definition.

First of all we say that if any fraction whatever be
taken, and if it be greater than the ratio of A C to A B, it
will also be greater than the ratio of A E to A D, and if

it be less than the one it will also be less than the other.

This is a matter which can be tested in regard to any particular fraction. If a length A E were given to us as the fourth proportional we could find out whether it obeyed the rule in respect of any one given fraction. But if there is a fourth proportional it must satisfy this rule in regard to all fractions whatever. We cannot directly test this; but we may be able to give a proof that the quantity which is supposed to be a fourth proportional obeys the rule for one particular fraction, which proof shall be applicable without change to any other fraction. It will then be proved, for this case, not only that a fourth proportional exists, but that this particular quantity is the fourth proportional. This is, in fact, just what we can do with the sides of similar

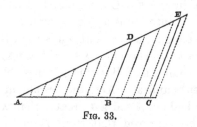

Fig. 33.

triangles. If the length A B (fig. 33) is divided into any number of equal parts, and lines are drawn through the points of division, making with A B the same angle that B D makes with it, they will divide A D into the same number of equal parts.

If now we set off points of division at the same distance from one another from B towards C, and through them draw lines making the same angle with the line A C that B D does, these lines will also cut off equal distances from D towards E. If any one of these lines starts from A C on the side of C towards

A, it will meet A E on the side of E towards A; because the triangle which it forms with the lines A C and A E must have the same shape as A C E. So also any one of these lines which starts from A C on the side of C away from A will meet A E on the side of E away from A.

Looking then at the various fractions of A B which are now marked off, it is clear that, if one of them is less than A C, the corresponding fraction of A D is less than A E; and if greater, greater. It follows, therefore, that the line A E which is given by this construction satisfies, in the case of any fraction we choose, the condition which is necessary for the fourth proportional. Consequently, if the second assumption which we made about space be true, there always is a fourth proportional, and this process will enable us to find it.

There is, however, still one objection to be made against our definition of the fourth proportional, or rather one point in which we can make it a firmer ground-work for the study of ratios. For it assumes that quantities are continuous; that is, that any quantity can be divided into any number of equal parts, this being implied in the process of taking any numerical fraction of a quantity.

We say, for example, that if a, b, c, d, are proportionals, and if a is greater than three-fifths of b, c will be greater than three-fifths of d. Now the process of finding three-fifths of b is one or other of the following two processes. Either we divide b into five equal parts and take three of them, or we multiply b by three and divide the result into five equal parts. (We know of course that these two processes give us the same result.) But it is assumed in both cases that we can divide a given quantity into five equal parts.

Now in a definition it is desirable to assume as

little as possible; and accordingly the Greek geometers
in defining proportion, or (which is really the same
thing) in defining the fourth proportional of three
given quantities, have tried to avoid this assumption.

Nor is it difficult to do this. For let us consider
the same example. We say that if a is greater than
three-fifths of b, c will be greater than the same fraction
of d. Now let us multiply both the quantities a and b
by five. Then for a to be greater than three-fifths of b,
the quantity which a has now become must be greater
than three-fifths of the quantity which b has become;
that is, if the new b be divided into five equal parts the
new a must be greater than three of them. But each of
these five equal parts is the same as the original b; and
so our statement as to the relative greatness of a and b
is the same as this, that five times a is greater than
three times b; and similarly for c and d.

Now every fraction involves two numbers. It is a
compound process made up of multiplying by one
number and dividing by another, and it is clear therefore that we may, not only in this particular case of
three-fifths but in general, transform our rule for the
fourth proportional into this new form. According as
m times a is greater or less than n times b, so is m times
c greater or less than n times d, where m and n are any
whole numbers whatever.

This last form is the one in which the rule is given
by the Greek geometers; and it is clear that it does
not depend on the continuity of the quantities considered, for whether it be true or not that we can
divide a number into any given number of equal parts,
we can certainly take any multiple of it that we
like.

These fundamental ideas, of ratio, of the equality of

ratios, and of the nature of the fourth proportional are now established generally, and with reference to quantities of any kind, not with regard to lengths alone ; provided merely that it is always possible to take any given multiple of any given quantity.

§ 6. *Of Areas; Stretch and Squeeze.*

We shall now proceed to apply these ideas to areas, or quantities of surface, and in particular to plane areas. The simplest of these for the purposes of measurement is a rectangle. The finding of the area of a rectangle is in many cases the same process as numerical multiplication. For example, a rectangle which is 7 inches long and 5 inches broad will contain 35 square inches, and this follows from our fundamental ideas about the multiplication of numbers. But this process, the multiplication of numbers, is only applicable to the case in which we know how many times each side of the rectangle contains the unit of length, and it then tells us how many times the area of the rectangle contains the square described upon the unit of length. It remains to find a method which can always be used.

For this purpose we first of all observe that when one side of a rectangle is lengthened or shortened in any ratio, the other side being kept of a fixed length, the area of the rectangle will be increased or diminished in exactly the same ratio.

In order then to make any rectangle O P R Q out of a square O A C B, we have first of all to stretch the side O A until it becomes equal to O P, and thereby to stretch the whole square into the rectangle O D, which increases its area in the ratio of O A to O P. Then we must stretch the side O B of this figure until it is equal to O Q, and

I

thereby the figure O D becomes O R, and its area is increased in the ratio of O B to O Q. Or we may, if we

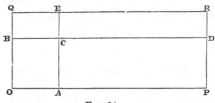

FIG. 34.

like, first stretch O B to the length O Q, whereby the square O C becomes O E, and then stretch O A to O P, by which O E becomes O R.

Thus the whole operation of turning the square O C into the rectangle O R is made up of two stretches; or, as we have agreed to call them, 'multiplications'; viz. the square has to be multiplied by the ratio of O P to O A, and by the ratio of O Q to O B; and we may find from the result that the order of these two processes is immaterial.

For let us represent the ratio of O P to O A by the letter a, and the ratio of O Q to O B by b. Then the ratio of the rectangle O D to the square O C is also a; in other words, a times O C is equal to O D. And the ratio of O R to O D is b, so that b times O D is equal to O R; that is, b times a times O C is equal to O R, or, as we write it, ba times O C is O R.[1]

And in the same way b times O C is equal to O E and a times b times O C is a times O E, which is O R.

---

[1] It is a matter of convention which has grown up in consequence of our ordinary habit of reading from left to right, that we always read the symbols of a multiplication, or of any other operation, *from right to left.* Thus $a\,b$ times any quantity x, means a times b times x; that is to say, we first multiply x by b, and then by a; that operation being first performed whose symbol comes last.

Consequently we have $b\,a$ times $o\,c$ giving the same result as $a\,b$ times $o\,c$; or, as we write it

$$b\,a = a\,b,$$

which means that the effect of multiplying first by the ratio a and then by the ratio b is the same as that of multiplying first by the ratio b and then by the ratio a.

This proposition, that in multiplying by ratios we may take them in any order we please without affecting the result, can be put into another form.

Suppose that we have four quantities, a, b, c, d, then I can make a into d by two processes performed in succession; namely, by first multiplying by the ratio of b to a, which turns it into b, and then by the ratio of d to b. But I might have produced the same effect on a by first multiplying it by the ratio of c to a, which turns it into c, and then multiplying by the ratio of d to c. We are accustomed to write the ratio of b to a in shorthand in any of the four following ways:—

$$b : a, \quad \frac{b}{a}, \quad b \div a, \quad {}^{b}/_{a},$$

and so the fact we have just stated may be written thus:—

$${}^{b}/_{a} \times {}^{d}/_{b} = {}^{c}/_{a} \times {}^{d}/_{c}.$$

Now let us assume that the four quantities, a, b, c, d, are proportionals; that is, that the ratios ${}^{b}/_{a}$ and ${}^{d}/_{c}$ are equal to one another. It follows then that the ratios ${}^{c}/_{a}$ and ${}^{d}/_{b}$ are equal to one another.

This proposition may be otherwise stated in this form; that if a, b, c, d are proportionals, then a, c, b, d will also be proportionals: provided always that this latter statement has any meaning, for it is quite possible that it should have no meaning at all. Suppose, for instance, that a and b are two lengths, c and d two intervals

of time, then we understand what is meant by the ratio
of b to a, and the ratio of d to c, and these ratios may
very well be equal to one another ; but there is no such
thing as a ratio of c to a, or of d to b, because the
quantities compared are not of the same kind. When,
however, four quantities *of the same kind* are proportionals, they are also proportionals when taken *alter-
nately* ; that is to say, when the two middle ones are
interchanged.

§ 7. *Of Fractions.*

We have seen in § 3, page 101, that a ratio may be
expressed in the form of a fraction. Thus, let a be
represented by the fraction $\dfrac{p}{q}$ and b by the fraction $\dfrac{r}{s}$,
where p, q, r, s are numbers. Then the result on page
115 may be written—

$$\frac{p}{q} \times \frac{r}{s} = \frac{r}{s} \times \frac{p}{q}.$$

Let us examine a little more closely into the meaning of either side of this equation. Suppose we were

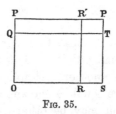

Fig. 35.

to take a rectangle O Q T S, of which one side, O Q, contained q units of length, and another, O S, s units.
Then this rectangle could be obtained from the unit
square by operating upon it with the two stretches q
and s. Its area would thus contain $q s$ square units.

Now let us apply to this rectangle in succession the two stretches denoted by $\frac{p}{q}$ and $\frac{r}{s}$. If we stretch the rectangle in the direction of the side o q in the ratio of $\frac{p}{q}$, we divide the side o q into q equal parts, and then take o p equal p times one of those parts. But each of these parts will be equal to unity, hence o p contains p units. We thus convert our rectangle o t into one o p′, of which one side, o p, contains p and the other, o s, s units. Now let us apply to this rectangle the stretch $\frac{r}{s}$ parallel to the side o s (as the figure is drawn

$\frac{r}{s}$ denotes a *squeeze*). We must divide o s into s equal parts and take r such parts, or we must measure a length o r along o s equal to r units. Thus this second stretch converts the rectangle o p′ into a rectangle o r′, of which the side o p contains p and the side o r contains r units of length, or into a rectangle containing $p\,r$ square units. Hence the two stretches $\frac{p}{q}$ and $\frac{r}{s}$ applied in succession to the rectangle o t convert it into the rectangle o r′. Now this may be written symbolically thus :—

$$\frac{p}{q} \times \frac{r}{s} . \text{ rectangle o t } = \text{ rectangle o r}'$$
$$= p\,r \text{ unit-rectangles.}$$

Now unit-rectangle may obviously be obtained from the rectangle o t by squeezing it first in the ratio $\frac{1}{q}$ in the direction of o q, and then in the ratio $\frac{1}{s}$ in the direction o s. Now this is simply saying that o t contains

$q\,s$ unit-rectangles. Hence the operation $\dfrac{p}{q} \times \dfrac{r}{s}$ applied to unit-rectangle must produce $\dfrac{1}{q\,s}$ of the result of its application to the rectangle o t. That is :—

$$\frac{p}{q} \times \frac{r}{s} \,.\, \text{unit-rectangle} = \frac{1}{q\,s} \,.\, p\,r \text{ unit-rectangle,}$$

or, in our notation, $= \dfrac{p\,r}{q\,s} \,.\, \text{unit-rectangle.}$

Hence we may say that $\dfrac{p}{q} \times \dfrac{r}{s}$ operating upon unity is equal to the operation denoted by $\dfrac{p\,r}{q\,s}$, or to multiplying unity by $p\,r$ and then dividing the result by $q\,s$. This equivalence is termed the *multiplication of fractions*.

A special case of the multiplication of fractions arises when s equals r. We then have—

$$\frac{p}{q} \times \frac{r}{r} = \frac{p\,r}{q\,r}.$$

But the operation $\dfrac{r}{r}$ denotes that we are to divide unity into r equal parts, and then take r of them ; in other words, we perform a *null* operation on unity. The symbol of operation may therefore be omitted, and we read—

$$\frac{p}{q} = \frac{p\,r}{q\,r}.$$

This result is then expressed in words as follows : Given a fraction, we do not alter its value by multiplying the numerator and denominator by equal quantities.

From this last result we can easily interpret the operation

$$\frac{p}{q} + \frac{r}{s}.$$

For, by the preceding paragraph—

$$\frac{p}{q} = \frac{ps}{qs}, \text{ and } \frac{r}{s} = \frac{qr}{qs}.$$

Hence—

$$\frac{p}{q} + \frac{r}{s} = \frac{ps}{qs} + \frac{qr}{qs}.$$

Or, to apply first the operation $\frac{p}{q}$ to unity and then to add to this the result of the operation $\frac{r}{s}$ is the same thing as dividing unity into qs parts, taking ps of those parts, and then adding to them qr more of the like parts. But this is the same thing as to take at once $ps + qr$ of those parts. Thus we may write—

$$\frac{p}{q} + \frac{r}{s} = \frac{ps + qr}{qs}.$$

This result is termed the *addition of fractions*. The reader will find no difficulty in interpreting addition graphically by a succession of stretches and squeezes of the unit-rectangle.

We term division the operation by which we reverse the result of multiplication. Hence when we ask the meaning of *dividing* by the fraction $\frac{p}{q}$ we put the question : What is the operation which, following on the operation $\frac{p}{q}$, just reverses its effect?

Now, $$\frac{r}{s} \times \frac{p}{q} = \frac{p}{q} \times \frac{r}{s} = \frac{pr}{qs}.$$

Suppose we take $r = q$, $s = p$.

Then $$\frac{q}{p} \times \frac{p}{q} = \frac{pq}{qp};$$

or, to multiply unity by $\frac{p}{q}$, and then by $\frac{q}{p}$, is to perform the operation of dividing unity into qp parts and then taking pq of them, or to leave unity unaltered. Hence the stretch $\frac{q}{p}$ completely reverses the stretch $\frac{p}{q}$; it is, in fact, a squeeze which just counteracts the preceding stretch. Thus multiplying by $\frac{q}{p}$ must be an operation equivalent to *dividing* by $\frac{p}{q}$. Or, to divide by $\frac{p}{q}$ is the same thing as to multiply by $\frac{q}{p}$. This result is termed the *division of fractions*.

§ 8. *Of Areas; Shear.*

Hitherto we have been concerned with stretching or squeezing the sides of a rectangle. These operations alter its area, but leave it still of rectangular shape. We shall now describe an operation which changes its angles, but leaves its area unaltered.

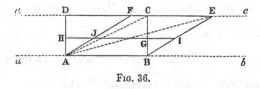

FIG. 36.

Let A B C D be a rectangle, and let A B E F be a parallelogram (or a four-sided figure whose opposite sides are equal), having the same side, A B, as the rectangle, but having the opposite side, E F (equal to A B, and

therefore to C D), somewhere in the same line as C D.
Then, since C D is equal to E F, the points E and F are
equally distant from C and D respectively, and it follows
that the triangles B C E and A D F are equal. Hence if
the triangle B C E were cut off the parallelogram along
B C and placed in the position A D F, we should have
converted the parallelogram into the rectangle without
changing its area. Thus the area of the parallelogram
is equal to that of the rectangle. Now the area of the
rectangle is the product of the numerical quantity which
represents the length of A D into that quantity which
represents the length of A B. A B is termed the *base*
of the parallelogram, and A D, the perpendicular distance between its base and the opposite side E F, is
termed its *height*. The area of the parallelogram is
then briefly said to be ' the product of its base into
its height.'

Suppose C D and A B were rigid rods capable of sliding along the parallel lines *c d* and *a b*. Let us imagine
them connected by a rectangular elastic membrane,
A B C D; then as the rods were moved along *a b* and *c d*
the membrane would change its shape. It would, however, always remain a parallelogram with a constant
base and height; hence its area would be unchanged.
Let the rod A B be held fixed in position, and the rod
C D pushed along *c d* to the position E F. Then any line,
G H, in the membrane parallel and equal to A B will be
moved parallel to itself into the position I J, and will
not change its length. The distance through which
C has moved is C E, and the distance through which G
has moved is G I. Since the triangles C B E and G B I
have their sides parallel they are similar, and we have
the ratio of C E to G I the same as that of B C to B G;
or, when the rectangle A B C D is converted into the

parallelogram A B E F, any line parallel to A B remains unchanged in length, and is moved parallel to itself through a distance proportional to its distance from A B. Such a transformation of figure is termed a *shear*, and we may consider either our rectangle as being sheared into the parallelogram or the latter as being sheared into the former. Thus the area of a parallelogram is equal to that of a rectangle into which it may be sheared.

The same process which converts the parallelogram A B E F into the rectangle A B C D will convert the triangle A B F, the half of the former, into the triangle

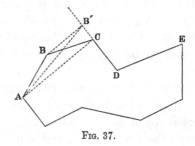

FIG. 37.

A B C, the half of the latter. Hence we may shear any triangle into a right-angled triangle, and this will not alter its area. Thus the area of any triangle is half the area of the rectangle on the same base, and with height equal to the perpendicular upon the base from the opposite angle. This height is also termed the altitude, or height of the triangle, and we then briefly say : *The area of a triangle is half the product of its base into its altitude.*

A succession of shears will enable us to reduce any figure bounded by straight lines to a triangle of equal area, and thus to determine the area the figure encloses by finally shearing this triangle into a right-angled triangle. For example, let A B C D E be a portion of the boundary of the figure. Suppose A C joined; then shear the triangle A B C so that its vertex B falls at B' on D C produced. The area A B' C is equal to the area A B C. Hence we may take A B' D E for the boundary of our figure instead of A B C D E; that is, we have reduced the number of sides in our figure by one. By a succession of shears, therefore, we can reduce any figure bounded by straight lines to a triangle, and so find its area.

§ 9. *Of Circles and their Areas.*

One of the first areas bounded by a curved line which suggests itself is that of a *sector* of a circle, or the

Fig. 38.

portion of a circle intercepted by two radii and the arc of the circumference between their extremities. Before we can consider the area of this sector it will be necessary to deduce some of the chief properties of the complete circle. Let us take a circle of unit radius and suppose straight lines drawn at the extremities of two diameters A B and C D at right angles; then the circle will appear as if drawn inside a square (see fig. 39). The sides of this square will be each 2 and its area 4.

Now suppose the figure composed of circle and square first to receive a stretch such that every line

parallel to the diameter A B is extended in the ratio of
$a : 1$, and then another stretch such that every line
parallel to C D is again extended in the ratio of $a : 1$.
Then it is obvious that we shall have stretched the
square of the first figure into a second square whose
sides will now be equal to $2a$.

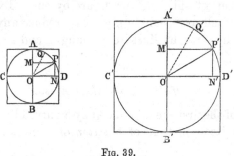

FIG. 39.

It remains to be shown that we have stretched the
first circle into another circle. Let O P be any radius
and P M, P N perpendiculars on the diameters A B, C D.
As a result of the first stretch the equal lengths O M
and N P are extended into the equal lengths O′ M′ and
N′ P′, which are such that $\dfrac{O\ M}{O'\ M'} = \dfrac{N\ P}{N'\ P'} = \dfrac{1}{a}$. Similarly
as a result of the second stretch M P and O N, which
remained unaltered during the first stretch, are con-
verted into M′ P′ and O′ N′; so that $\dfrac{O\ N}{O'\ N'} = \dfrac{M\ P}{M'\ P'} = \dfrac{1}{a}$.
During this second stretch O′ M′ and N′ P′ remain un-
altered. Thus as the total outcome of the two stretches
we find that the triangle O P N has been changed into the
triangle O′ P′ N′. Now these two triangles are of the
same shape by what was said on p. 106, for the angles
at N and N′ are equal, being both right angles, and we
have seen that—

$$\frac{N\ P}{N'\ P'} = \frac{1}{a} = \frac{O\ M}{O'\ M'}.$$

Thus it follows that the third side O P must be to the third side O' P' in the ratio of 1 to a; or, since O P is of unit length, O' P' must be equal to the constant quantity a. Further, since the angles P O N, P' O' N' are equal, O' P' is parallel to O P. Hence the circle of unit radius has been stretched into a circle of radius a. In fact, the two equal stretches in directions at right angles, which we have given to the first figure, have performed just the same operation upon it, as if we had placed it under a magnifying glass which enlarged it uniformly, and to such a degree that every line in it was magnified in the ratio of a to 1.

It follows from this that the circumference of the second circle must be to that of the first as a is to 1. Or, the circumferences of circles are as their radii. Again, if the arc P Q is stretched into the arc P' Q'—that is, if O' P', O' Q' are respectively parallel to O P, O Q—then the arc P' Q' is to the arc P Q in the ratio of the radii of the two circles. Since the arcs P Q, P' Q' are equal to any other arcs which subtend the same angles at the centres of their respective circles, we state generally that *the arcs of two circles which subtend equal angles at their respective centres are in the ratio of the corresponding radii.*

Since the second figure is an uniformly magnified image of the first, every element of area in the first has been magnified at the same uniform rate in the second. Now the square in the first figure contains four units of area, and in the second figure it contains $4\,a^2$ units of area. Hence every element of area in the first figure has been magnified in the second in the ratio of a^2 to 1. Thus the area of the circle in the first figure must be to the area of the circle in the second figure as 1 is to a^2. Or : *The areas of circles are as the squares of their radii.*

It is usual to represent the area of a circle of unit radius by the quantity π; thus the area of a circle of radius a will be represented by the quantity πa^2.

If, after stretching A B to A′ B′ in the ratio of a to 1, we had stretched or squeezed C D to C′ D′ in the ratio of b to 1, where b is some quantity different from a, our square would have become a rectangle, with sides equal to $2\,a$ and $2\,b$ respectively. It may be shown that we

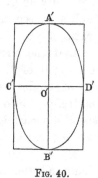

FIG. 40.

should have distorted our circle into the shape of that shadow of a circle which we have termed an ellipse. Furthermore, elements of area have now been stretched in the ratio of the product of a and b to 1 ; or, the area of the ellipse is to the area of the circle of unit radius as $a\,b$ is to 1: whence it follows that the area of the ellipse is represented by $\pi\,a\,b$, where a and b are its greatest and least radii respectively.

We shall now endeavour to connect the area of a circle of unit radius, which we have written π, with the number of linear units in its circumference. Let us take a number of points uniformly distributed round the circumference of a circle, A B C D E F. Join them in succession to each other and to o, the centre of the circle, and draw the lines perpendicular to these radii (or the *tangents*) at A B C D E F; then we shall have constructed two perfectly symmetrical figures, one of which is said to be *inscribed*, the other *circumscribed* to the circle. Now the areas of these two figures differ by the sum of such triangles as A *a* B, and the area of the circle is obviously greater than the area of the inscribed and less than the area of the circumscribed figure. Thus

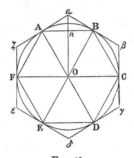

Fig. 41.

the area of the circle must differ from that of the inscribed figure by something less than the sum of all the little triangles A *a* B, B β C, &c. Now from symmetry all these little triangles are equal, and their areas are therefore equal to one half the product of their heights, or *a n*, into their bases, or such quantities as A B. Hence the sum of their areas is equal to one half of the product of *a n* into the sum of the sides of the inscribed figure. Now the sum of the sides of the inscribed figure is never greater than the circumference of the circle. If we take, therefore, a great number of points uniformly distributed round the circumference of our circle, A and

B may be brought as close as we please, and the nearer we bring A to B, the smaller becomes $a\,n$. Hence, by taking a sufficient number of points, we can make the sum of the triangles A a B, B β C, &c. as small as we please, or the areas of the inscribed and circumscribed figures, together with the area of the circle which lies between them, can be made to differ by less than any assignable quantity. In the limit then we may say that by taking an indefinite number of points we can make these areas equal. Now the area of the inscribed figure is the sum of the areas of all such triangles as A O B, and the area of the triangle A O B is equal to half the product of its height O n into its base A B ; or if we write for the 'perimeter,' or sum of all the sides A B, B C, &c. the quantity p, the area of the inscribed figure will equal $\frac{1}{2}\,p \times$ O n. Again if p' be the sum of the sides $a\,\beta$, $\beta\,\gamma$, &c. of the circumscribed figure, its area $= \frac{1}{2}\,p' \times$ O B.

Since the triangles O a B, O B n are of the same shape, being right-angled and again equi-angled at O, we have the ratio of B n to a B, or of their doubles A B to $a\,\beta$, the same as that of O n to O B. But p is obviously to p' in the same ratio as A B to $a\,\beta$; hence p is to p' as O n to O B. By taking a sufficient number of points we can make O n as nearly equal to O B as we please ; thus we can make p as nearly equal to p', and therefore either of them as nearly equal to the circumference of the circle (which lies between them),[1] as we please. Hence in the limit p will equal the circumference of the circle, and O n its radius, and we may state that the areas of the inscribed and circumscribed figures, which approach nearer and nearer to the area of the circle as we increase the number of their sides, become ultimately

---

[1] In the case of the circle the reader will recognise this intuitively.

equal to each other and to half the product of the circumference of the circle into its radius. This must therefore be the area of the circle. Hence we have the following equality:—The area of a circle of radius a equals
one half its circumference × a. But it equals also πa^2;
whence it follows that the circumference of a circle
equals $\pi . 2\, a$. We may express this result in two
different ways:—

(i) The ratio of the circumference of a circle to its
diameter $(2\, a)$ is a constant quantity π.

(ii) The number of linear units $(2\,\pi)$ in the circumference of a circle of unit-radius is twice the
number of units of area (π) contained by that circumference.

The value of π, the ratio of the circumference of a
circle to its diameter, is found to be a quantity which,
like the ratio of the diagonal of a square to its side (see
p. 103), cannot be expressed accurately by numbers;
its approximate value is 3·14159.

We have now no difficulty in finding the area of
the sector of a circle, for if we double the arc of a
sector we obviously double its area; if we treble it, we
treble its area; shortly, if we take any multiple of it,
we take the same multiple of its area. Hence it
follows by § 5, that two sectors are to each other
in the ratio of their arcs, or a sector must be to the
whole circle in the ratio of its arc to the whole circumference.

If we represent by s the area of a sector of a circle
of which the arc contains s units of length and the
radius a units, we may write this relation symbolically—

$$\frac{s}{\pi a^2} = \frac{s}{2\,\pi a}.$$

K

Thus we deduce $s = \frac{1}{2} s \times a$; or,
*The area of a sector is half the product of the length of its
arc into its radius.*

§ 10. *Of the Area of Sectors of Curves.*

The knowledge of the area of a sector of a circle
enables us to find as accurately as we please the area
of a sector whose arc is any curve whatever. Let the
arc P Q be divided into a number of smaller arcs P A, A B,
B C, C D, D Q. We shall suppose that P A subtends the
greatest angle at O of all these arcs. Further we shall
consider only the case where the line O P diminishes
continuously if P be made to pass along the arc from P

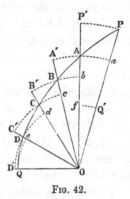

FIG. 42.

to Q. If this be not the case, the sector Q O P can
always be split up into smaller sectors, of which it shall
be true that a line drawn from the point O to the arc continuously diminishes from one side of the sector to the
other, and then for the area of each of these sectors the
following investigation will hold. With O as centre describe a circle of radius O P to meet O A produced in P′; with
the same centre and radius O A describe a circle to meet

O B in A′ and O P in a; similarly circles with radius O B to meet O A in b and O C in B′, with radius O C to meet O B in c and O D in C′, with radius O D to meet O C in d and O Q in D′, and finally with radius O Q to meet O D in e, O A in f, and O P in Q′. Then the area of the sector obviously lies between the areas of the figure bounded by O P, O D′ and the broken line P P′ A A′ B B′ C C′ D D′, and of the figure bounded by O a, O Q and the broken line a A b B c C d D e Q. Hence it differs from either of them by less than their difference or by less than the sum of the areas P′ a, A′ b, B′ c, C′ d, D′ e. Now since the angle at P O P′ is greater than any of the other sectorial angles at O, the sum of all these areas must be less than that of the figure P P′ f Q′, and the area of this figure can be made as small as we please by making the angle A O P sufficiently small. This can be achieved by taking a sufficient number of points like A, B, C, D, &c. We are thus able to find a series of circular sectors, the sum of whose areas differs by as small a quantity as we please from the area of the sector P O Q; in other words, we reduce the problem of finding the area of any figure bounded by a curved line to the problem already solved of finding the area of a sector of a circle. The difficulties which then arise are purely those of adding together a very great number of quantities; for, it may be necessary to take a very great number of points such as A B C D . . . in order to approach with sufficient accuracy to the magnitude of the area P O Q.

§ 11. *Extension of the Conception of Area.*

Let A B C D be a closed curve or loop, and O a point inside it. Then if a point P move round the perimeter of the loop, the line O P is said to trace out the area of

the loop A B C D. By this is meant that successive positions of the line o P, pair and pair, form together with
the intervening elements of arc elementary sectors, the
sum of the areas of which can, by taking the successive

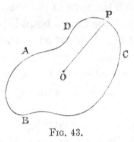

Fig. 43.

positions sufficiently close, be made to differ as little as
we please from the area bounded by the loop.

Now suppose the point o to be taken *outside* the
loop A B C D, and let us endeavour to find the area then

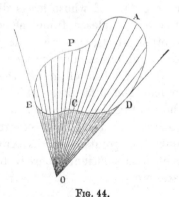

Fig. 44.

traced out by the line o P joining o to a point P which
moves round the loop. Let o B and o D be the extreme
positions of the line o P to the left and to the right as
P moves round the loop A B C D; then as P moves along

the portion of the loop D A B, O P moves *counter-clock-wise* from right to left and traces out the area bounded by the arc D A B and the lines O D and O B. Further, as P moves along the portion of the loop B C D, O P moves *clockwise* from left to right and traces out the area doubly shaded in our figure, or the area bounded by the arc B C D and the lines O B and O D. It is the *difference* of these two areas which is the area of the loop A B C D. If, then, we were to consider the latter area O B C D O as *negative*, the line O P would still trace out the area of the loop A B C D as P moves round its perimeter. Now the characteristic difference in the method of describing the areas O D A B O and O B C D O is, that in the former case O P moves *counter-clockwise* round O, in the latter case it moves *clockwise*. Hence if we make a convention that areas traced out by O P when it is moving counter-clockwise shall be considered positive, but areas traced out by O P when it is moving clockwise shall be considered negative, then wherever O may be inside or outside the loop, the line O P will trace out its area provided P move completely round its circumference.

But it must here be noted that P may describe the loop in two different methods, either going round it counter-clockwise in the order of points A B C D, or clockwise in the order of points A D C B. In the former case, according to our convention, the greater area O D A B O is positive, in the latter it is negative. Hence we arrive at the conception that *an area may have a sign*; it will be considered positive or negative according as its perimeter is supposed traced out by a point moving counter-clockwise or clockwise. This extended conception of area, as having not only magnitude but *sense*, is of fundamental importance, not only in many

branches of the exact sciences, but also for its many
practical applications.[1]

Let a perpendicular o n be erected at o (which is,
as we have seen, any point in the plane of the loop)
to the plane of the loop, and let the length o n be
taken along it containing as many units of length as
there are units of area in the loop A B C D. Then o n
will represent the area of the loop in magnitude; it
will also represent it in *sense*, if we agree that o n shall
always be measured in such a direction from o, that to
a person standing with his feet at o and head at n the
point P shall always appear to move counter-clockwise.
Thus, for a positive area, n will be above the plane;
for a negative area, in the opposite direction or below
the plane. We are now able to represent any number
of areas by segments of straight lines or steps perpendicular to their planes. The sum of any number of
areas lying in the same plane will then be obtained by
adding algebraically all the lines which represent these
areas.

When the areas do not all lie in one plane the
representative lines will not all be parallel. In this
case there are two methods of adding areas. We may
want to know the total amount of area, as, for example,
when we wish to find the cost of painting or gilding
a many-sided solid. In this case we add all the representative lines without regard to their direction.

In many other cases, however, we wish to find some
quantity so related to the sides of a solid that it can
only be found by treating the lines which represent
their areas as *directed* magnitudes. Such cases, for
example, arise in the discussion of the shadows cast by

[1] As in calculating the cost of levelling and embanking, in the indicator
diagram, &c. It was first introduced by Möbius.

the sun or of the pressure of gases upon the sides of a containing vessel, &c. A method of combining directed magnitudes will be fully discussed in the following chapter. The conception of areas as directed magnitudes is due to Hayward.

§ 12. *On the Area of a Closed Tangle.*

Hitherto we have supposed the areas we have talked about to be bounded by a simple loop. It is easy, however, to determine the area of a combination of loops. Thus consider the figure of eight in fig. 45 which has two loops : if we go round it continuously in the direction indicated by the arrow-heads, one of these loops will have a positive, the other a negative area, and therefore the total area will be their difference, or zero if they be equal. When a closed curve, like a figure of eight, cuts itself it is termed a *tangle*, and the points where it cuts itself are called *knots*. Thus a figure of eight is a tangle of one knot. In tracing out the area of a closed curve by means of a line drawn from a fixed point to a point moving round the curve, the area may vary according to the direction and the route by which we suppose the curve to be described. If, however, we suppose the curve to be sketched out by the moving point, then its area will be perfectly definite for that particular description of its perimeter.

We shall now show how the most complex tangle may be split up into simple loops and its whole area determined from the areas of the simple loops. We shall suppose arrow-heads to denote the direction in which the perimeter is to be taken. Consider either of the accompanying figures. The moving line o p will trace out exactly the same area if we suppose it not to cross at the knot A but first to trace out the loop A C and then to trace out the loop A B, in both cases going round these two loops in the direction

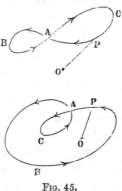

FIG. 45.

indicated by the arrow-heads. We are thus able in all cases to convert one line cutting itself in a knot into two lines, each bounding a separate loop, which just touch at the point indicated by the former knot. This dissolution of knots may be suggested to the reader by leaving a vacant space where the boundaries of the loops really meet. The two knots in the following figure are shown dissolved in this fashion:—

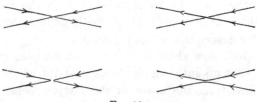

FIG. 46.

The reader will now find no difficulty in separating the most complex tangle into simple loops. The positive or negative character of the areas of these loops will be sufficiently indicated by the arrow-heads on their perimeters. We append an example :—

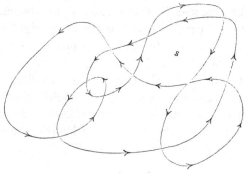

Fig. 47.

In this case the tangle reduces to a negative loop a, and to a large positive loop b, within which are two other positive loops c and d, the former of which con-

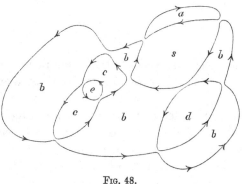

Fig. 48.

tains a fifth small positive loop e. The area of the entire tangle then equals $b + c + d + e - a$. The space marked s in the first figure will be seen from the second to be no part of the area of the tangle at all.

§ 13. *On the Volumes of Space-Figures.*

Let us consider first the space-figure bounded by
three pairs of parallel planes mutually at right angles.
Such a space-figure is technically termed a 'rectangular
parallelepiped,' but might perhaps be more shortly
described as a 'right six-face.' We may first observe
that when one edge of such a right six-face is
lengthened or shortened in any ratio, the other non-
parallel edges being kept of a fixed length, the volume

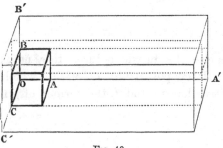

Fig. 49.

will be increased in precisely the same ratio. Hence,
in order to make any right six-face out of a cube we
have only to give the cube three stretches (or it may
be squeezes), parallel respectively to its three sets of
parallel edges. Let o A, o B, o C be the three edges of
the cube which meet in a corner o. Let o A be
stretched to o A', so that the ratio of o A' to o A is
represented by a; then if the figure is to remain right
all lines parallel to o A will be stretched in the same
ratio. The figure has now become a six-face whose
section perpendicular to o A' only is a square. Now
stretch o B to o B', so that the ratio o B' to o B be
represented by b, and let all lines parallel to o B be

increased in the same ratio ; the figure is now a right
six-face, only one set of edges of which are equal to the
edge of the original square. Finally stretch o c to o c′,
so that o c and all lines parallel to it are increased in
the ratio of o c′ to o c, which we will represent by c.
By a process consisting of three stretches we have thus
converted our original cube into a right six-face. If
the cube had been of unit-volume, the volume of our
six-edge would obviously be abc, and we may show as
in the case of a rectangle (see p. 115) that $abc = cba$
$= bac$, &c. ; or the order of multiplying together three
ratios is indifferent. If we term the face A′ c′ of our

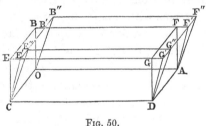

Fig. 50.

right six-face its *base* and o b′ its *height*, a c will represent the area of its base, and b its height, or the volume
of a right six-face is equal to the product of its base
into its height.

Let us now suppose a right six-face o A D C E B F G
to receive a shear, or the face B E F G to be moved in its
own plane in such fashion that its sides remain parallel
to their old positions, and B and E move respectively
along B F and E G. If B′ E′ G′ F′ be the new position of
the face B E G F, it is easy to see that the two wedge-
shaped figures B E E′ B′ O C and F G G′ F′ A D are exactly
equal; this follows from the equality of their corresponding faces. Hence the volume of the sheared

figure must be equal to the volume of the right six-face. Now let us suppose in addition that the face B′ E′ G′ F′ is again moved in its own plane into the position B″ E″ G″ F″, so that B′ and E′ move along B′ E′ and F′ G′ respectively. Then the slant wedge-shaped figures B′ B″ F″ F′ A O and E′ E″ G″ F′ D C will again be equal, and the volume of the six-face B″ E″ G″ F″ A D C O obtained by this second shear will be equal to the volume of the figure obtained by the first shear, and therefore to the volume of the right six-face. But by means of two shears we can move the face B E G F to any position in its plane, B″ E″ G″ F″, in which its sides remain parallel to their former position. Hence the volume of a six-face will remain unchanged if, one of its faces, O C D A, remaining fixed, the opposite face, B E G F, be moved anywhere parallel to itself in its own plane. We thus find that the volume of a six-face formed by three pairs of parallel planes is equal to the product of the area of one of its faces and the perpendicular distance between that face and its parallel. For this is the volume of the right six-face into which it may be sheared; and, as we have seen, shear does not alter volume.

The knowledge thus gained of the volume of a six-face bounded by three pairs of parallel faces, or of a so-called parallelepiped, enables us to find the volume of an *oblique cylinder*. A right cylinder is the figure generated by any area moving parallel to itself in such wise that any point P moves along a line P P′ at right angles to the area. The volume of a right cylinder is the product of its height P P′ and the generating area. For we may suppose that volume to be the sum of a number of elementary right six-faces whose bases, as at P, may be taken so small that they will ultimately completely fill the area A C B D, and whose heights are all equal to P P'.

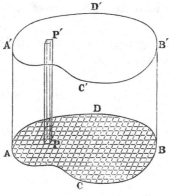

FIG. 51.

We obtain an oblique cylinder from the above right cylinder by moving the face A' C' B' D' parallel to itself anywhere in its own plane. But such a motion will only shear the elementary right six-faces, such as P P', and so not change their volume. Hence the volume of an oblique cylinder is equal to the product of its base, and the perpendicular distance between its faces.

§ 14. *On the Measurement of Angles.*

Hitherto we have been concerned with quantities of area and quantities of volume; we must now turn to quantities of *angle*. In our chapter on Space (p. 66) we have noted one method of measuring angles; but that was a merely relative method, and did not lead us to fix upon an absolute unit. We might, in fact, have taken any opening of the compasses for unit angle, and determined the magnitude of any other angle by its ratio to this angle. But there is an absolute unit which naturally suggests itself in our measurement of angles, and one which we must consider here, as we shall frequently have to make use of it in our chapter on Position.

Let A O B be any angle, and let a circle of radius a be described about O as centre to meet the sides of this

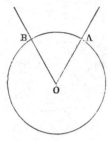

Fig. 52.

angle in A and B. Then if we were to double the angle A O B, we should double the arc A B; if we were to treble it, we should treble thé arc; shortly, if we were to take any multiple of the angle, we should take the same multiple of the arc. We may thus state that angles at the centre of a circle vary as the arcs on which they stand. Hence if θ and θ' be two angles, which are subtended by arcs s and s' respectively, the ratio of θ to θ' will be the same as that of s to s'. Now suppose θ' to represent four right angles; then s' will be the entire circumference, or, in our previous notation, $2\pi a$. We have thus—

$$\frac{\theta}{\text{four right angles}} = \frac{s}{2\pi a}.$$

Now it is extremely convenient to choose a unit angle which shall be independent of the circle upon which we measure our arcs. We should obtain such an independent unit if we took the arc subtended by it equal to the radius of the circle or if we took $s = a$. In this case our unit equals $\dfrac{1}{2\,\pi}$ of four right angles,

$= \dfrac{1}{\pi}$ of two right angles, $= \cdot636$ of a right angle approximately.

Thus we see that the angle subtended at the centre of any circle by an arc equal to the radius is a constant fraction of a right angle.

If this angle be chosen as the unit, we deduce from the proportion θ is to θ' as s is to s', that θ must be to unity as s is to the radius a; or :—

$$s = a\,\theta.$$

Thus, if we choose the above angle as our unit of angle, the measure of any other angle will be the ratio of the arc it subtends from the centre to the radius; but we have seen (p. 125) that the arcs subtended from the centre in different circles by equal angles are in the ratio of the radii of the respective circles. Hence the above measurement of angle is *independent of the radius of the circle upon which we base our measurement.* This is the primary property of the so-called *circular measurement* of angles, and it is this which renders it of such great value.

The *circular measure* of any angle is thus the ratio of the arc it subtends from the centre of any circle to the radius of the circle. It follows that the circular measure of four right angles is the ratio of the whole circumference to the radius, or equals $\dfrac{2\,\pi\,a}{a}$; that is, equals $2\,\pi$. The circular measure of two right angles will then be π, of one right angle $\dfrac{\pi}{2}$, of three right angles $\dfrac{3\,\pi}{2}$, and so on.

§ 15. *On Fractional Powers.*

Before we leave the subject of quantity it will be necessary to refer once more to the subject of powers which we touched upon in our chapter on Number (p. 16).

We there used a^n as a symbol signifying the result of multiplying a by itself n times. From this definition we easily deduce the following identity :—

$$a^n \times a^p \times a^q \times a^r = a^{n+p+q+r}.$$

For the left hand side denotes that we are first to multiply a by itself n times, and then multiply this by a^p, or a multiplied by itself p times, and so on. Hence we may write the left hand side—

$$(a \times a \times a \times a \ .. \text{ to } n \text{ factors})$$
$$\times (a \times a \times a \times a \ .. \text{ to } p \text{ factors})$$
$$\times (a \times a \times a \times a \ .. \text{ to } q \text{ factors})$$
$$\times (a \times a \times a \times a \ .. \text{ to } r \text{ factors}).$$

But this is obviously equal to $(a \times a \times a \times a \times \ ...$ to $n+p+q+r$ factors), or to $a^{n+p+q+r}$.

If b be such a quantity that $b^n = a$, b is termed an nth *root* of a, and this is written symbolically $b = \sqrt[n]{a}$. Thus, since $8 = 2^3$, 2 is a 3rd, or cube root of 8. Or, again, since $243 = 3^5$, 3 is termed a 5th root of 243.

Now we have seen at the conclusion of our first chapter that we can often learn a very great deal by extending the meaning of our terms. Let us now see if we cannot extend the meaning of the symbol a^n. Does it cease to have a meaning when n is a fraction or negative? Obviously we cannot multiply a quantity by itself a fractional number of times, nor can we do

so a negative number of times.   Hence the old meaning of a^n, where n is a positive integer, becomes sheer
nonsense when we try to adapt it to the case of n
being fractional or negative.   Is then a^n in this latter
case meaningless?

In an instance like this we are thrown back upon
the results of our definition, and we endeavour to give
to our symbol such a meaning that it will satisfy these
results.   Now the fundamental result of our theory of
integer powers is that—

$$a^{n+p+q+r+\cdots} = a^n \times a^p \times a^q \times a^r \times \ldots$$

This will obviously be true however many quantities,
n, p, q, r, we take.   Now let us suppose we wish to interpret $a^{\frac{l}{m}}$ where $\frac{l}{m}$ is a fraction.   We begin by assuming it satisfies the above relation, and in order to
arrive at its meaning we suppose that $n = p = q$
$= r = \ldots = \frac{l}{m}$, and that there are m such quantities.
Then

$$n + p + q + r = m \times \frac{l}{m} = l;$$

and we find $a^l = a^{\frac{l}{m}} \times a^{\frac{l}{m}} \times a^{\frac{l}{m}} \times \ldots$ to m factors

$$= \left(a^{\frac{l}{m}}\right)^m.$$

Thus $a^{\frac{l}{m}}$ must be such a quantity that, multiplied by
itself m times, it equals a^l.   But we have defined above
(p. 144) an mth root of a^l to be such a quantity that,
multiplied m times by itself, it equals a^l.   Hence we
say that $a^{\frac{l}{m}}$ is equal to an mth root of a^l; or, as it is
written for shortness,—

$$a^{\frac{l}{m}} = \sqrt[m]{a^l}.$$

L

We have thus found a meaning for a^n when n is a fraction from the fundamental theorem of powers.

We can with equal ease obtain from the same theorem an intelligible meaning for a^n when n is a negative quantity. We have $a^n \times a^p = a^{n+}$. Now let us assume $p = -n$ in order to interpret a^{-n}. We find $a^n \times a^{-n} = a^{n-n} = a^0 = 1$ (by p. 31). Or dividing by a^n,

$$a^{-n} = \frac{1}{a^n};$$

that is to say, a^{-n} is the quantity which, multiplied by a^n, gives a product equal to unity. The former quantity is termed the *inverse* of the latter, or we may say that a^{-n} is the inverse of a^n. For example, what is the inverse of 4? Obviously 4 must be multiplied by $\frac{1}{4}$ in order that the product may be unity. Hence 4^{-1} is equal to $\frac{1}{4}$. Or, again, since $4 = 2^2$, we may say that 2^{-2} is the *inverse* of 4, or 2^2.

The whole subject of powers—integer, fractional, and negative—is termed the *Theory of Indices*, and is of no small importance in the mathematical investigation of symbolic quantity. Its discussion would, however, lead us too far beyond our present limits. It has been slightly considered here in order that the reader may grasp that portion of the following chapter in which fractional powers are made use of.

CHAPTER IV.

POSITION.

§ 1. *All Position is Relative.*

THE reader can hardly fail to remember instances when he has been accosted by a stranger with some such question as : 'Can you tell me where the 'George' Inn lies ? '—' How shall I get to the cathedral ? '—' Where is the London Road ? ' The answer to the question, however it may be expressed, can be summed up in the one word—*There.* The answer points out the *position* of the building or street which is sought. Practically the *there* is conveyed in some such phrase as the following : 'You must keep straight on and take the first turning to the right, then the second to the left, and you will find the ' George ' two hundred yards down the street.'

Let us examine somewhat closely such a question and answer. 'Where is the ' George ' ? ' We may expand this into : ' How shall I get from *here*' (the point at which the question is asked) 'to the ' George ' ? ' This is obviously the real meaning of the query. If the stranger were told that the ' George ' lies three hundred paces from the Town Hall down the High Street, the information would be valueless to the questioner unless he were acquainted with the position of the Town Hall or at least of the High Street. Equally idle

L 2

would be the reply: 'The 'George' lies just past the forty second milestone on the London Road,' supposing him ignorant of the whereabouts of the London Road.

Yet both these statements are in a certain sense answers to the question: 'Where is the 'George'?' They would be the true method of pointing out the *there*, if the question had been asked in sight of the Town Hall or upon the London Road. We see, then, that the query, *Where*? admits of an infinite number of answers according to the infinite number of positions—or possible *heres*—of the questioner. The *where* always supposes a definite *here*, from which the desired position is to be determined. The reader will at once recognise that to ask, 'Where is the 'George'?' without meaning, 'Where is it with regard to some other place?' is a question which no more admits of an answer than this one: 'How shall I get from the 'George' to anywhere?' meaning to nowhere in particular.

This leads us to our first general statement with regard to position. We can only describe the *where* of a place or object by describing how we can get at it from some other known place or object. We determine its *where* relative to a *here*. This is shortly expressed by saying that: All position is relative.

Just as the 'George' has only position relative to the other buildings in the town, or the town itself relative to other towns, so a body in space has only position relative to other bodies in space. To speak of the position of the earth in space is meaningless unless we are thinking at the same time of the Sun or of Jupiter, or of a star—that is, of some one or other of the celestial bodies. This result is sometimes

described as the ' sameness of space.' By this we only
mean that in space itself there is nothing perceptible to
the senses which can determine position.[1] Space is, as
it were, a blank map into which we put our objects; it
is the coexistence of objects in this map which enables
us at any instant to distinguish one object from another.
This process of distinguishing, which supposes at least
two objects to be distinguished, is really determining
a *this* and a *that*, a *here* and a *there*; it involves the
conception of relativity of position.

§ 2. *Position may be Determined by Directed Steps.*

Let us turn from the question: 'Where is the
' George '?' to the answer : ' You must keep straight on
and take the first turning to the right, then the second
to the left, and you will find the 'George' 200 yards
down the street.'

The instruction ' to keep straight on' means to keep
in the street wherein the question has been asked, and
in a direction ('straight on') suggested by the previous
motion of the questioner, or by a wave of the hand from
the questioned.   Assuming for our present purpose
that the streets are not curved, this amounts to: Keep
a certain direction.   How far? This is answered by the
second instruction: Take the first turning on the right.
More accurately we might say, if the first turning to the
right were 150 yards distant: Keep this direction for
150 yards.   Let this be represented in our figure by the
step A B, where A is the position at which the question
is asked.   At B the questioner is to turn to the right
and, according to the third instruction, he is to pass the
first turning to the left at c and take the second at D.

---

[1] We shall return to this point later.

More accurately we might state the distance B D to be, say, 180 yards.   Then we could combine our second and third instructions by saying: From B go 180 yards in a certain direction, namely, B D.   To determine exactly what this direction B D is with regard to the first direction A B, we might use the following method.   If the stranger did not change his direction at B, but went straight on for 180 yards, he would come to a point D'.   Hence if we measured the angle D'B D between the street in which the question was asked and the first turning to the right,

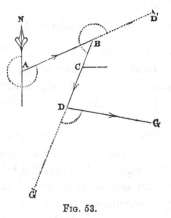

Fig. 53.

we should know the direction of B D and the position of D exactly.   It would be determined by rotating B D' about B through the measured angle D'B D.   If we adopt the same convention for the measurement of positive angles as we adopted for positive areas on p. 133, the angle D'B D is the angle greater than two right angles through which B D' must be rotated counter-clockwise in order to take it to the position B D.   Let us term this angle D'B D for shortness β, then we may invent a new symbol $\{\beta\}$ to denote the operation : Turn the direction you are going in through an angle β counter-clockwise.

If we use the symbol $\pi/2$ to denote an angle equal to a right angle, we have the following symbolic instructions:

$\{\ 0\ \}\ =$ Keep straight on.

$\{\ \pi/2\ \}\ =$ Turn at right angles to the left.

$\{\ \pi\ \}\ =$ Turn right round and go back.

$\{3\pi/2\}\ =$ Turn at right angles to the right.

Thus for a turning from A B to the left the angle of our symbolic operation will be less, for a turning from A B to the right greater, than two right angles.

If the directed person had gone to D′ instead of to D, he would have walked 150 yards to B and then 180 yards to D′; he would thus have walked A B + B D′, or 150 yards + 180 yards. In order to denote that he is not to continue straight on at B we introduce the operator of turning, namely $\{\beta\}$, before the 180 yards, and read $150 + \{\beta\}180$ as the instruction: Go 150 yards along some direction A B, and then, turning your direction through an angle β counter-clockwise, go 180 yards along this new direction.

We are now able to complete the symbolic expression of our instructions for finding the 'George.' The fourth instruction runs: Take a turning at D to the left and go 200 yards along the direction thus determined. Let D G′ represent 200 yards measured from D along B D produced, then we are to revolve D G′ through a certain angle G′D G counter-clockwise, till it takes up the position D G. Then G will be the position of the 'George.' Let the angle G′D G be represented by γ. Our final instruction may be then expressed symbolically by $\{\gamma\}200$.

Hence our total instruction may be written symbolically—

$$150 + \{\beta\}180 + \{\gamma\}200,$$

where the units are yards.

But we have not yet quite freed this symbolic instruction from any suggestion of direction as determined by streets; the first 150 yards are still to be taken along the *street* in which the question is asked. We can get rid of this street by supposing its direction determined by the angle which a clock-hand must revolve through counter-clockwise, to reach that direction, starting from some other fixed or chosen direction. For example, suppose the stranger to have a compass with him, and at A let A N be the direction of its needle. Then we might fix the position of the street A B by describing it as a direction so many degrees east of north, or still to preserve our counter-clockwise method of reckoning angles, we might determine it by the angle a which the needle would have to describe through west and south to reach the position A B. We should then interpret the notation $\{a\}150$: Walk 150 yards along a direction making an angle a with north measured through west.

Our answer expressed symbolically is now entirely cleared of any conception of streets. For,

$$\{a\}150 + \{\beta\}180 + \{\gamma\}200$$

is a definite instruction as to how to get from A to G quite independent of any local characteristics. It expresses the position of G with regard to A in a purely geometrical fashion, or by a series of *directed steps*. Expanded into ordinary English our symbols read: From a point A in a plane, take a step A B of 150 units in a direction making an angle a with a fixed direction, from B take a step B D of 180 units making an angle β with A B, and finally from D take a step D G of 200 units making an angle γ with B D. All the angles are to be measured counter-clockwise in the fashion we have described above.

§ 3. *The Addition of Directed Steps or Vectors.*

If we now compare our figure with the symbolical instruction $\{a\}\,150 + \{\beta\}\,180 + \{\gamma\}\,200$, we see that $\{a\}$ 150 represents the step A B, when that step is considered to have not merely magnitude but also *direction*. Similarly B D and D G represent more than linear expressions for number—they are also *directed* steps. We shall then be at liberty to replace our symbolically expressed instruction

$$\{a\}\,150 + \{\beta\}\,180 + \{\gamma\}\,200$$

by the geometrical equivalent

$$AB + BD + DG,$$

provided we understand by the segments A B, B D, D G and the symbol + something quite different to our

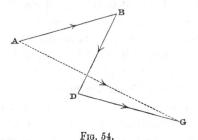

Fig. 54.

former conceptions. We give a new and extended meaning to our quantity and to our addition.

A B + B D + D G no longer directs us to add the number of units in B D to that in A B and to the sum of these the number in D G, but it bids us take a step A B in a certain *direction*, then a step B D from the finish of the former step in another determined direction, and finally from the finish D of this second step a third

directed step, D G.   The entire operation brings us
from A to G.   Now it is obvious that we should also
have got to G had we taken the directed step A G.
Hence, if we give an extended meaning to the word
'equal' and to its sign =, using them to mark the
equivalence of the results of two operations, we may
write

$$A G = A B + B D + D G,$$

and read this expression :—A G equals the sum of A B,
B D and D G.

Steps such as we considered in our chapter on
Quantity, which were magnitudes taken along any one
straight line, are termed *scalar* steps, because they have
relation only to some chosen scale of quantity.   We
add or subtract scalar steps by placing them end to end
in *any* straight line (see § 2 of Chapter III.)

A step which has not only magnitude but *direction*
is termed a *vector* step, because it *carries* us from one
position in space to another.   It is usual to mark by an
arrow-head the sense in which we are to take this
directed step.   For example in fig. 54 we are to step
from A to B, and thus the arrow-head will point towards
B for the step A B.   In letters this is denoted by writing
A before B.   The method by which we have arrived at
the conception of vector steps shows us at once how to
add them.

Vector steps are added by placing them end to end
in such fashion that they retain their own peculiar
directions, and so that a point moving continuously
along the zigzag thus formed will always follow the
directions indicated by the arrow-heads.   This may be
shortly expressed by saying the steps are to be arranged
in *continuous sense*.   The sum of the vector steps is
then the single directed step which joins the start of

the zigzag thus formed to its finish. In fig. 55 let *a b*, *c d*, *e f*, and *g h* be directed steps. Then let A B be drawn equal and parallel to *a b*; from B draw B C equal and parallel to *c d*, from C draw C D equal and parallel to *e f*, and finally from D draw D E equal and parallel to *g h*. We have drawn our zigzag so that the arrow-heads all have 'a continuous sense.' Hence the directed step A E is the sum of the four given vectors. If, for example, at C we had stepped C D′, equal and parallel to *e f*, but on the opposite side of B C to C D, and then taken D′E′,

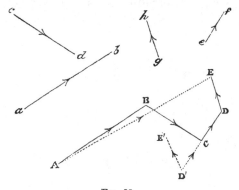

FIG. 55.

equal and parallel to *g h*, the reader will remark at once that the arrow-heads in B C, C D′ and D′E′ are not in continuous sense, or we have not gone in the proper direction at C.

Should the vector steps all have the same direction, the zigzag evidently becomes a straight line; in this case the vector steps are added precisely like scalar quantities; or, when vector steps may be looked upon as scalar, our extended conception of addition takes the ordinary arithmetical meaning.

We can now state a very important aspect of position

in a plane; namely, if the position of G relative to A be denoted by the directed step or vector A G, it may also be expressed by the sum of any number of directed steps, the start of the first of such steps being at A and the finish of the last at G (see fig. 56). We may write this result symbolically :—

$$A G = A B + B C + C D + D E + E F + F G.$$

It will be at once obvious that in our example as to finding the 'George,' the stranger might have been directed by an entirely different set of instructions to

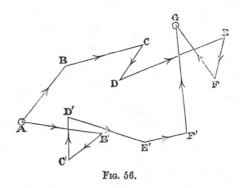

Fig. 56.

his goal. In fact, he might have been led to make extensive circuits in or about the town before he reached the place he was seeking. But, however he might get to G, the ultimate result of his wanderings would be what he might have accomplished by the directed step A G supposing no obstacles to have been in his way (or, 'as the crow flies'). Hence we see that with our extended conception of addition any two zigzags of directed steps, A B C D E F G and A B′ C′ D′ E′ F′ G (which may or may not contain the same number of component steps), both starting in A and finishing in G, must be looked upon as equivalent instructions; or, we must take

$$A B + B C + C D + D E + E F + F G = A G =$$
$$A B' + B'C' + C'D' + D'E' + E'F' + F'G.$$

In other words, two sets of directed steps must be held to have an equal sum, when, their starts being the same, the steps of both sets will, added vector-wise have the same finish.

Now let us suppose our stranger were unconsciously standing in front of the 'George' when he asked his question as to its whereabouts, and further let us suppose that the person who directed him gave him a perfectly correct instruction, but sent him by a properly chosen set of right and left turnings a considerable distance round the town before bringing him back to the point A from which he had set out. In this case we must suppose the 'George' not to be at the point G, but at the point A. The total result of the stranger's wanderings having brought him back to the place from which he started can be denoted by a zero step; or we must write (fig. 56)—

$$A B + B C + C D + D E + E F + F G + G A = 0 \ . \ . \ . \text{(i)}$$

We may read this in words: The sum of vector steps which form the successive sides of a closed zigzag is zero. Now we have found above that—

$$A B + B C + C D + D E + E F + F G = A G \ . \ . \ . \ . \ . \text{(ii)}$$

Hence, in order that these two statements (i) and (ii) may be consistent, we must have — G A equal to A G, or

$$A G + G A = 0.$$

This is really no more than saying that if a step be taken from A to G, followed by another from G to A, the total operation will be a zero step. Yet the result is

interesting as showing that if we consider a step from
A to G as positive, a step from G to A must be considered
negative.   It enables us also to reduce subtraction of
vectors to addition.   For if we term the operation
denoted by A B − D C a *subtraction* of the vectors A B
and D C, since D C + C D = 0, the operation indicated
amounts to adding the vectors A B and C D, or to
A B + C D.   Hence, to subtract two vectors, we reverse
the sense of one of them and add.

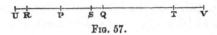

FIG. 57.

The result A G + G A = 0 can at once be extended to
any number of points lying on a straight line.   Thus, if
P Q R S T U V be a set of such points—

$$P Q + Q R + R S + S T + T U + U V + V P = 0.$$

For starting from P and taking in succession the steps
indicated, we obviously come back to P, or have performed an operation whose result is equivalent to zero,
or to remaining where we started.

§ 4.   *The Addition of Vectors obeys the Commutative Law.*

We can now prove that the commutative law holds
for our extended addition (see p. 5).   First, we can
show that any two successive steps may be interchanged.
Consider four successive steps, A B, B C, C D, and D E.
If at B instead of taking the step B C we took a step
B H equal to C D in magnitude, sense and direction, we
could then get from H to D by taking the step H D.
Now let B D be joined; then in the triangles B H D, D C B
the angles at B and D are equal, because they are formed
by the straight line B D falling on two parallel lines B H

and C D ; also the side B D is common, and B H is equal
to C D. Hence it follows (see p. 73) that these triangles
are of the same shape and size, or H D is equal to B C ;
and again the angles B D H and D B C are equal, or H D
and B C are parallel. Thus the step H D is equal to the
step B C in direction, magnitude and sense. We have
then from the two methods of reaching D from B,

$$B\,C + C\,D = B\,D = B\,H + H\,D$$
$$= C\,D + B\,C$$

by what we have just proved.

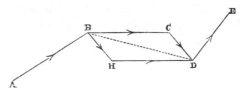

Fig. 58.

Hence any two successive steps may be interchanged. By precisely the same reasoning as we have
used on p. 11 we can show that if we may interchange any two successive steps of our zigzag we may
interchange any two steps whatever by a series of
changes of successive steps; that is, the order in
which vectors are added is indifferent.

The importance of the geometry of vectors arises
from the fact that many physical quantities can be represented as directed steps. We shall see in the succeeding chapter that velocities and accelerations are
quantities of this character.

§ 5. On Methods of Determining Position in a Plane.

It has been remarked (see p. 99) that scalar
quantities may be treated as steps measured along a

straight line. In this case we only require one point on this line to be given, and we can determine the relative position of any other by merely stating the magnitude of the intervening step. A line is occasionally spoken of as being a space of *one* dimension; in one-dimensioned space one point suffices to determine the relative position of all others.

When we consider however position in a plane, in order to determine the whereabouts of a point P with regard to another A we require to know not only the magnitude but the *direction* of the step A P. Hence what scalar steps are to one-dimensioned space, that

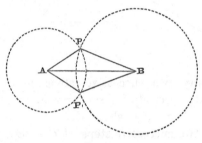

are vector steps to plane space. In order to determine the *direction* of a step A P we must know at least one other point B in the plane. Space which requires two points to determine the position of a third is usually termed space of *two* dimensions. There are various methods in general use by which position in two-dimensioned space is determined. We shall mention a few of them, confining our remarks however to the plane, or to space of two dimensions which is of the same shape on both sides.

(*a*) We may measure the distances between A and P and between B and P. If these distances are of scalar magnitude r and r' respectively, there will be two points corresponding to any two given values of r and r'; namely P and P' the intersections of the two circles with centres at A and B and radii equal to r and r' respectively. We may distinguish these points as being one above, and the other below A B. Only in the case of the circles touching will the two points coincide; if the circles do not meet, there will be no point.

If P moves so that for each of its positions with regard to A and B the quantities r and r' satisfy some definite relation, we shall obtain a continuous set of points in the plane or a curved line of some sort. For example, if we fasten the ends of a bit of string of length l to

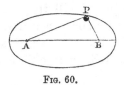

FIG. 60.

pins stuck into the plane of the paper at A and B, and then move a pencil about so that its point P always remains on the paper, and at the same time always keeps the string A P B taut round its point, the pencil will trace out that shadow of the circle which we have called an ellipse.

In this case $r + r' = \text{A P} + \text{P B} = l$, the constant length of the string. This relation $r + r' = l$ is an equation between the scalar quantities r, r' and l, which holds for every point on the ellipse, and expresses a metric property of the curve with regard to the points A and B.

If on the other hand we cause P to move so that the difference of A P and B P is a constant length $(r - r' = l)$, then P will trace out the curve we have termed the

M

hyperbola. We can cause P to move in this fashion by means of a very simple bit of mechanism. Suppose a rod B L capable of revolving about one of its ends B : let a string of given length be fastened to the other end L and to the fixed point A. Then if, as the rod is moved round B, the string be held taut to the rod by a

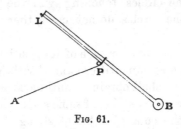

FIG. 61.

pencil point P, the pencil will trace out the hyperbola. For since L P + P A equals a constant length, namely that of the string, and L P + P B equals a constant length, namely that of the rod, their difference or P A − P B is equal to the constant length which is the difference of the string and the rod.

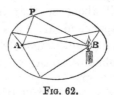

FIG. 62.

The points A and B are termed in the cases of both ellipse and hyperbola the *foci*. The name arises from the following interesting property. Suppose a bit of polished watch spring were bent into the form of an ellipse so that its flat side was turned towards the foci of the ellipse; then if a hot body were placed at one focus B, all the rays of heat or light radiated from B which fell upon the spring would be collected, or, as it is termed, 'focussed' at A; hence A would be a much brighter and hotter point than any other within the ellipse (B of course excepted). The name *focus* is from the Latin, and means a fireplace or hearth. This property of the arc of an ellipse or hyperbola, that it collects rays radiating from one focus in the other, depends upon the fact that A P and P B make equal angles with the curve at P. This geometrical relation corresponds to a physical property of rays of heat and light; namely, that they make the same angle with a reflecting surface when they reach it and when they leave it.

A third remarkable curve, which is easily obtained from this our first method of considering position, is the lemniscate of James Bernoulli (from the Latin *lemniscus*, a ribbon). It is traced out by a point P which moves so that the rectangle under its distances from A and B is always equal to the area of a given square

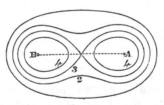

FIG. 63

$(r . r' = c^2)$. If the given square is greater than the square on half A B, it is obvious that P can never cross between A and B; if it is equal to the square on half A B, the lemniscate becomes a figure of eight; while if it is less, the curve breaks up into two loops. In our figure a series of lemniscates are represented. A set of curves obtained by varying a constant, like the

M 2

given square in the case of the lemniscate, is termed a *family of curves*. Such families of curves constantly occur in the consideration of physical problems.

§ 6. *Polar Co-ordinates.*

(β) The points A and B determine a line whose direction is A B. If we know the length A P and the angle B A P, we shall have a means of finding the position of P. Let r be the number of linear units in A P and θ the number of angular units in B A P, where r and θ may of course be fractions. In measuring the angle θ we shall adopt the same convention as we have employed in discussing areas (see p. 134) ; namely, if a line at first coincident with A B were to start from

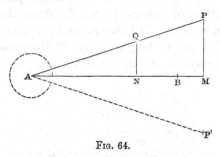

FIG. 64.

that position, and supposed pivoted at A to rotate counter-clockwise till it coincided with A P, it would trace out the angle θ. Angles traced out clockwise will like areas be considered negative. Thus the angle B A P′ below A B would be obtained by a rotation clockwise from A B to A P′, and must therefore be treated as negative. On the other hand, we might have caused a line rotating about A to take up the position A P′ by rotating it counter-clockwise through an angle marked in our figure by the dotted arc of a circle. Further we might obviously have reached A P by a line rotating about A clockwise, and might thus represent the position of P by a negative angle. But even after we had got to P we might cause our line to rotate about A a complete number of times either clockwise or counter-clockwise, and we should still be at the end of any such number of complete revolutions in the same position A P.

We have then the following four methods of rotating a line about A from coincidence with A B to coincidence with A P : —

(i) Counter-clockwise from A B to A P.

(ii) Clockwise from A B to A P.

(iii) The first of these combined with any number of complete revolutions clockwise or counter-clockwise.

(iv) The second of these combined with any number of complete revolutions clockwise or counter-clockwise.

The following terms have been adopted for this method of determining position in space :—

The line A B from which we begin to rotate our line is termed the *initial* (' beginning ') line ; the length A P is termed the *radius vector* (from two Latin words signifying the carrying rod or spoke, because it carries the point P to the required position) ; the angle B A P is termed the *vectorial angle*, because it is traced out by the radius vector in moving from A B to the required position A P ; A is termed the *pole*, because it is the end of the axis about which we may suppose the spoke to turn. Finally A P $(= r)$ and the angle B A P $(= \theta)$ are termed the *polar co-ordinates* of the point P, because they regulate the position of P relative to the pole A and the initial line A B.

§ 7. *The Trigonometrical Ratios.*

If P M be a perpendicular dropped from P on A B, the ratios of the sides of the right-angled triangle P A M have for the purpose of abbreviation been given the following names :—

$\frac{PM}{AP}$, or the ratio of the perpendicular to the hypothenuse, is termed the *sine* of the angle B A P.

$\frac{AM}{AP}$, or the ratio of the base to the hypothenuse, is termed the *cosine* of the angle B A P.

$\frac{PM}{AM}$, or the ratio of the perpendicular to the base, is termed the *tangent* of the angle B A P.

$\frac{AM}{PM}$, or the ratio of the base to the perpendicular, is termed the *cotangent* of the angle B A P.

If θ be the scalar magnitude of the angle B A P these ratios are written for shortness, $sin\theta$, $cos\theta$, $tan\theta$, and $cot\theta$, respectively. Let us take any other point Q on A P, and drop Q N perpendicular to A B, then the triangles Q A N, P A M are of the same shape (see p. 106), and thus the ratios of their corresponding sides are equal. It follows from this that the ratios sine, cosine, tangent, and cotangent for the triangles Q A N and P A M are the same. Hence we see that $sin\theta$, $cos\theta$, $tan\theta$, and $cot\theta$ are independent of the position of P in A P; they are ratios which depend only on the magnitude of the *angle* B A P or θ. They are termed (from two Greek words meaning *triangle-measurement*) the trigonometrical ratios of the angle θ. The discussion of trigonometrical ratios, or *Trigonometry*, forms an important element of pure mathematics. The names of the trigonometrical ratios themselves are derived from an older terminology which connected these ratios with the figure supposed to be presented by an archer whose bow string was placed against his breast.[1]

§ 8. *Spirals.*

Let us suppose the spoke A P to revolve about the pole A, and as it revolves let the point P move along the spoke in such fashion that the magnitude r of A P is always definitely related in some chosen manner to the magnitude θ of B A P. Then if P be taken as the point of a pencil it will mark out a curved line on the plane of the paper.

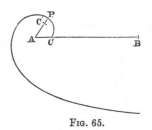

Fig. 65.

Such a curved line is termed a *polar curve* or *spiral*, the latter name from a Greek word denoting the coil, as of a snake, to which some of these curves may be considered to bear resemblance.

One of the most interesting of these spirals was invented by Conon of Samos (*fl.* B.C. 250), but its

---

[1] In our figure the angle B A P has been taken *less* than a right angle, it may have any magnitude whatever. It has been found useful to establish a convention with regard to the *signs* of the perpendicular P M and the base A M. P M is considered positive when it falls above, but negative when it falls below the initial line A B ; A M is considered positive when M falls to the right, but negative when it falls to the left of A. The reader will understand the value of this convention better after examining §§ 11, 12.

chief properties having been discussed by Archimedes,
it is usually called by his name. The spiral of Archimedes is defined in the following simple manner. As
the spoke A P moves uniformly round the pole, the point
P moves uniformly along the spoke. Let C be the position of P when the spoke coincides with the starting
line A B, and let A C contain a units of length. Then if P
be the position of the pencil-point when the spoke has
described an angle B A P containing θ units of angle, and
if A C' be measured along A P equal to A C, the point will
have described the distance C' P while the spoke was
turning through the angle C A P. But since the point
and spoke are moving uniformly, the distance C' P must
be proportional to the angle C A P, or their ratio must
be an unchangeable quantity for all distances and
angles. Let b be the distance traversed by the point
along the spoke while it turns through unit angle,
then C' P must be equal to the number of units in C A P
multiplied by b. Using r to denote the magnitude of
A P we have

$$C' P = b \times \theta, \text{ but } C' P = r - a;$$

Thus : $r = a + b\,\theta.$

This relation between r and θ is termed the polar *equation* to the spiral.

The following easily constructed apparatus will
enable us to draw a spiral of Archimedes. D E F is a
circular disc of chosen radius ; upon the edge of this
disc is cut a groove. To the centre A of the disc is
attached a rod or spoke which can be revolved about A
as a pole ; at the other end of this rod is a small grooved
wheel or pulley G. A string is then fastened to some
point D in the groove of the disc, and passing round
the pulley G is attached to a small block P which holds

a pencil and is capable of sliding in a slot in the spoke. If this block be fastened by a piece of elastic to A, the string from P to G and then from G to the groove on the disc will remain taut. Now supposing the disc to be held firmly pressed against the paper, and the spoke A C to be turned about A counter-clockwise, the pencil P will describe the required spiral. For the string touching the disc in the point T the figure G A T always remains of the same size and shape as we turn the spoke about the pole; hence the length of string G T is

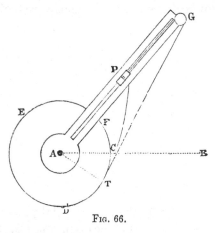

Fig. 66.

constant. Thus if a length of string represented by the arc D T be wound on to the disc as we turn the spoke from the position A B to the position A P, the length P G (since the length G T always remains the same) must lose a length equal to D T as P moves from C to P. But the amount of string D T wound on to the disc is proportional to the angle through which the spoke A P has been turned; hence the point P must have moved towards G through a distance proportional to this angle, or it has described a spiral of Archimedes.

Once in possession of a good spiral of this kind we can solve a problem which often occurs, namely to divide an angle into any number of parts having given ratios. Let the given angle be placed with its vertex at the pole of the spiral and let the radii vectores A C and A P be those which coincide with the legs of the angle. About the pole A describe a circular arc with radius A C to meet A P in C'. Now let us suppose the problem solved and let the radii vectores A D, A E, A F be those which divide the angle into the required proportional parts. If these radii vectores meet the circular arc C C' in D', E',

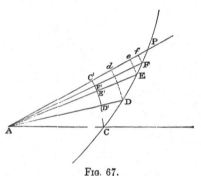

Fig. 67.

F' respectively, then by the fundamental property of the spiral we have at once the lines D'D, E'E, F'F, C'P in the same ratio as the angles C A D, C A E, C A F, C A P. Thus if we measure lengths A d, A e, A f equal to A D, A E, A F respectively along A P, C' P will be divided in d e f into lengths which are proportional to the required angles. Conversely, if we were to divide C'P into segments C'd, d e, e f, and f P in the same ratio as the required angular division, we should obtain lengths A d, A e, A f, which would be the radii of circles with a common centre A cutting the spiral in the required points of angular division. The spiral of Archimedes thus enables us to reduce the division of an angle in any fashion to the like division of a line.

Now the division of a line in any fashion, that is, into a set of segments in any given ratio, is at once solved so soon as we have learnt by the aid of a pair of compasses or a ' set square ' to draw parallel lines. Thus suppose we require to divide the line c′p into segments in the ratio of 3 to 5 to 4 ; we have only to mark off along any line through c′, say c′q, steps c′r, r s, s t placed end to end and containing 3, 5, and 4 units of any kind respectively. If the finish of the last step t be joined to p

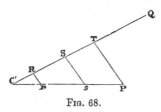

Fig. 68.

and the parallels r*r*, s*s* to t p through r and s be drawn to meet c′p in r and s, then c′p will be divided in r and s into segments in the required ratio of 3 to 5 to 4. This follows at once from our theory of triangles of the same shape (see p. 106). For, since r c′ r, s c′s, and t c′p are such triangles, they have their corresponding sides proportional, and the truth of the proposition is obvious.

A spiral of Archimedes accurately cut in a metal or ivory plate is an extremely useful addition to the ordinary contents of a box of so-called mathematical instruments.

§ 9. *The Equiangular Spiral.*

Another important spiral was invented by Descartes, and is termed from two of its chief properties either the *equiangular* or the *logarithmic spiral.*

Let B O A be a triangle with a small angle at O, and whose sides O A and O B are of any not very greatly different lengths. Upon O B and upon the opposite side of it to A construct a triangle B O C of the same shape as the triangle A O B, and in such wise that the angles at B and A are equal. Then upon O C place a triangle C O D of the same shape as either B O C or A O B; upon O D a fourth triangle D O E, again of the same shape; upon O E a fifth triangle, and so on. We thus ultimately form a figure consisting of a number of triangles A O B, B O C, C O D,

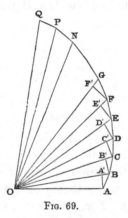

Fig. 69.

D O E, &c., of the same shape, all placed with one of their equal angles at O, and in such fashion that each pair has a common side consisting of two non-corresponding sides (that is, of sides not opposite to equal angles). The points A B C D E, &c., will form the angles of a polygonal line, and if the angles at O are only taken small enough, the sides of this polygon will appear to form a continuous curved line. This curved line, to which we can approach as closely as we please by taking the angles at O smaller and smaller, is termed an *equiangular spiral*. It derives its name from the following property,—A B, B C, C D, &c.,

being corresponding sides of triangles of the same shape, make equal angles o B A, o C B, o D C, &c., with the corresponding sides o B, o C, o D, &c. ; but when the angles at o are taken very small A B, B C, C D, &c., will appear as successive elements of the curved line or spiral. Hence the arc of the spiral meets all rays from the pole o at the same constant angle.

Let us now endeavour to find the relation between any radius vector o P ($=r$) and the vectorial angle A o P ($=\theta$).

Since all our triangles A o B, B o C, C o D, &c., are of the same shape, their corresponding sides must be proportional (see p. 106); or,

$$\frac{OB}{OA} = \frac{OC}{OB} = \frac{OD}{OC} = \frac{OE}{OD} = \frac{OF}{OE} = \&c.$$

Each of these equal ratios will therefore have the same scalar value; let us denote that value by the symbol μ. Then we must have

$$O B = \mu . O A ; \quad O C = \mu . O B ; \quad O D = \mu . O C ; \quad \&c.$$

Or, o B $= \mu . o A$; o C $= \mu^2 . o A$; o D $= \mu^3 . o A$, and so on. Hence if o N be the radius vector which occurs after n equal angles are taken at o, we must have

$$O N = \mu^n . O A.$$

Now let the very small angles at o be each taken equal to some small part of the unit angle ; thus we might take them $\frac{1}{100}$ or $\frac{1}{1000}$ of the unit angle. We will represent this fraction of the unit angle by $1/b$, where we may suppose b a whole number for greater simplicity. Further let us use λ to denote the bth power of μ, or $\lambda = \mu^b$. With the notation explained on p. 144 we then term μ a bth root of λ, and write $\mu = \lambda^{1/b}$.

Hence finally we have $O N = O A . \lambda^{n \times 1/b}$, or in words : The base of the nth equal-shaped triangle placed about o is equal to the base of the first multiplied by a certain quantity λ raised to the power of n-times the quantity $1/b$ which expresses the magnitude of the equal angles at o in units of angle.

Now let the spoke or ray O P fall within the angle which is formed by the successive rays O N and O Q of the system of equal-shaped triangles round o.   Then O N makes an angle n-times $1/b$, and O Q an angle $(n+1)$-times $1/b$ with O A.   Hence the angle A O P, or θ, must lie in magnitude between n/b and $(n+1)/b$.   Similarly the magnitude of O P must lie between those of O N and O Q. Now by sufficiently decreasing the angles at o we can approach nearer and nearer to the form of the spiral, and the ray O P must always lie between two successive rays of our system of triangles.   The angle θ, which will thus always lie between n/b and $(n+1)/b$, can only differ from either of them by a quantity less than $1/b$. If then b be taken large enough, or the equal angles at o small enough fractions of the unit angle, this difference $1/b$ can be made vanishingly small.   In this case we may say that *in the limit* the angle θ becomes equal to n/b and the ray O P equal to O N or O Q, which will thus be ultimately equal.   Hence O P $=$ O A . $\lambda^{n/b}=$ O A . λ^{θ}, or in words : If a ray O P of the equiangular spiral make an angle A O P with another ray O A, the ratio of O P to O A is equal to a certain number λ raised to the power of the quantity θ which expresses the magnitude of the angle A O P in units of angle.

If a and r be the numbers which express the magnitudes of O A and O P, we have $r = a \lambda^{\theta}$.   This is termed the *polar equation* of the spiral.

We proceed to draw some important results from a

consideration of this spiral. The reader will at once observe that the ratio of any pair of rays O P and O Q is equal to the ratio of any other pair which include an equal angle, for the ratio of any pair of rays depends only on the included angle. Further, if we wanted to multiply the ratio of any two quantities p and q by the ratio of two other quantities r and s we might proceed as follows : Find rays of the equiangular spiral O P, O Q, O R, O S containing the same number of linear units as p, q, r, s contain units of quantity (see p. 99), and let

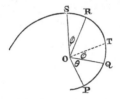

FIG. 70.

θ be the angle between the first pair, ϕ the angle between the second pair.

Then

$$\frac{OQ}{OP} = \lambda^\theta, \text{ and } \frac{OS}{OR} = \lambda^\phi ;$$

whence it follows that $\dfrac{OQ}{OP} \times \dfrac{OS}{OR} = \lambda^\theta \times \lambda^\phi = \lambda^{\theta+\phi}$, or is equal to the ratio of any pair of rays which include an angle $\theta + \phi$. Thus if the angle Q O T be taken equal to ϕ, and O T be the corresponding ray of the spiral, $\dfrac{OT}{OP} = \lambda^{\theta+\phi}$, and is a ratio equal to the product of the given ratios. Hence to find the product of ratios we have only to add the angles between pairs of rays in the given ratios, and the ratio of any two rays including an angle equal to the sum will be equal to the required product.   Thus the equiangular spiral enables us to *replace multiplication by addition*.   This is an extremely valuable substitution, as it is much easier to add than to multiply.

Since $\dfrac{\text{O Q}}{\text{O P}}$ divided by $\dfrac{\text{O S}}{\text{O R}} = \lambda^{\theta}$ divided by $\lambda^{\phi} = \lambda^{\theta - \phi}$,

it is obvious that we may in like fashion replace the division of two ratios by the subtraction of two angles.   A set of quantities like the angles at the pole of an equiangular spiral which enables us to replace multiplication and division by addition and subtraction is termed a *table of logarithms*.   Since the equiangular spiral acts as a graphical table of logarithms, it is frequently termed the *logarithmic spiral*.

§ 10.   *On the Nature of Logarithms.*

Since in the logarithmic spiral O P $=$ O A $\times \lambda^{\theta}$, where θ is equal to the angle A O P, we note that as θ increases, or as the ray O P revolves round O, O P is equally multiplied during equal increments of the vectorial angle A O P.   When one quantity depends upon another in such fashion that the first is equally multiplied for equal increments of the second, it is said to *grow at logarithmic rate*.   This logarithmic rate is measured by the ratio of the growth of the first quantity for unit increment of the second quantity to the magnitude of the first quantity before it started this growth.

Let us endeavour to apply this to our equiangular spiral.   Suppose A O B, B O C, C O D &c. to be as before the triangles by means of which we construct it (see fig. 69), the angles at O being all equal and very small. Along O B measure a length O A′ equal to O A; along O C, a length O B′ equal to O B; along O D, a length O C′

equal to o c, and so on. Then A'B, B'C, C'D, &c., will be the successive growths as a ray is turned successively from o A to o B, from o B to o C, and so on. Join A A', B B', C C', &c. Now the triangles A O B, B O C, C O D, &c., are all of the same shape; so too are the isosceles triangles A O A', B O B', C O C', &c. Hence the differences of the corresponding members of these sets, A A'B, B B'C, C C'D, &c., must also be of equal shape, and thus their corresponding sides proportional. It follows then that the lengths

A'B, B'C, C'D, &c., are in the same ratio as the lengths
A'A, B'B, C'C, &c., or again as the lengths
OA, OB, OC, &c.

Whence we deduce that

$$\frac{A'B}{O A} = \frac{B'C}{O B} = \frac{C'D}{O C} = \&c.$$

Or, the growth A'B is always in a constant ratio to the growing quantity o A.

Now, if the angles at o be very small, the line A A' will practically coincide with the arc of a circle with centre o and radius equal to o A. Hence (see p. 143) A A' will ultimately equal o A × the angle A O A', while the angle at A' will ultimately be equal to a right angle.

Further, the ratio of A'B to A A' remains the same for all the little triangles A A'B, B B'C, C C'D, &c. It is in each case the ratio of the *base* to the *perpendicular* when we look upon these triangles with regard to the equal angles A B A', B C B', C D C', &c. Now these are the angles of the triangles which give the spiral its name. Let any one of them, and therefore all of them, be equal to *a*. By definition the cotangent of an angle (see p. 166) is equal to the ratio of the base to the perpendicular.

N

Hence

$$\cot a = \frac{A'B}{A\,A'} = \frac{A'B}{OA \times \text{angle } A\,O\,A'},$$

or
$$\frac{A'B}{OA} = \text{angle } A\,O\,A' \times \cot a.$$

Now A B denotes the growth for an angle A O A', supposed very small; whence it follows that the *logarithmic rate*, or the ratio of the growth to the growing quantity for *unit* angle, is equal to cot$a$. Thus the logarithmic rate for the growth of the ray of the equiangular or logarithmic spiral, as it describes equal angles about the pole, is equal to the cotangent of the angle which gives its name to the spiral.

Let us suppose O A to be unit of length, then, since O P = O A × λ^θ, the result O P of revolving the ray O A through an angle θ equal to unity will be λ, or λ is the result of making unity grow at logarithmic rate cot$a$.

Now let us denote by the symbol e the result of making unity grow at logarithmic rate unity during the description of unit angle. Then e will have some definite numerical value. This value is found, by a process of calculation into which we cannot enter here, to be nearly equal to 2·718. This means that, if while unit ray were turned through unit angle it grew at logarithmic rate unity, its total growth (1·718) would lie between eight and nine-fifths of its initial length. Since e is the result of turning unit ray through unit angle, and since the ray is equally multiplied for equal multiples of angle, e^γ must represent the result of turning unit ray through γ unit angles. Hitherto we have been concerned with unit ray growing at logarithmic rate unity; now let us suppose unity to grow at logarithmic rate γ; then it grows γ times as much as if it grew at logarithmic rate unity, or the result of turning unit ray

through unit angle, while it grows at logarithmic rate
γ, must be the same as if we spread $1/\gamma$ of this rate
of growth over γ unit angles; that is, as if we caused
unity to grow at logarithmic unity for γ unit angles,
or e^γ. Hence e^γ denotes the result of making unit ray
grow at logarithmic rate unity while it describes γ unit
angles, or again of making unit ray grow at logarithmic rate γ while it describes a unit of angle.

Let us inquire what is the meaning of e^γ when γ is
a commensurable fraction equal to s/t, s and t being
integers. Let x be the as yet unknown result of turning unit ray through an angle equal to γ while it
grows at unit logarithmic rate; then x^t will be the
result of turning unit ray through t angles equal to
γ while it grows at unit rate; but t angles equal to γ
form an angle containing s units, or this result must
be the same as the result of turning unity through an
angle s while it grows at logarithmic rate t. Thus we
have $x^t = e^s$. That is, x is a t-th root of e^s, or, as we write
it, equal to $e^{s/t} = e^\gamma$. Thus e^γ, if γ be a commensurable
fraction, is the result of causing unit ray to grow at
logarithmic rate unity through an angle equal to γ, or
as we have seen at logarithmic rate γ through unit
angle.

Now let us suppose it possible to find a commensurable fraction γ equal to $\cot a$; then the result of
making unity grow at logarithmic rate $\cot a$ as it is
turned through unit angle must be e^γ. But we have
seen (see p. 178) that it is equal to λ. Hence

$$\lambda = e^\gamma.$$

Further, the result of making unity grow at logarithmic rate $\cot a$ as it is turned through an angle θ
is λ^θ; or,

$$\lambda^\theta = e^{\gamma\theta}.$$

Thus we may write $OP = OA \cdot \lambda^\theta = OA \cdot e^{\gamma\theta}$,
or with our previous symbols,

$$r = a \cdot e^{\gamma\theta}.$$

This is therefore the equation to our equiangular spiral expressed in terms of the quantity e.

If we take a spiral in which a is the unit of length, and in which $\cot a$ or γ is also unity, we find

$$r = e^\theta.$$

The symbol e^θ is then read the *exponential* of θ, and θ is termed the *natural logarithm* of r. It is denoted symbolically thus :—

$$\theta = \log_e r.$$

The quantity e is termed the *base* of the natural system of logarithms. Our spiral would in this case form a graphical table of *natural logarithms*.

Returning to the equation

$$r = a \cdot e^{\gamma\theta},$$

let us suppose γ so chosen that $e^\gamma = 10$; then γ will represent the angle through which unit ray must be turned in order that, growing at unit logarithmic rate, it may increase to ten units. Again taking a to be of unit length we find $r = e^{\gamma\theta} = 10^\theta$. θ is in this case termed the logarithm of r to the base 10, and this is symbolically expressed thus :—

$$\theta = \log_{10} r.$$

The spiral obtained in this case would form a graphical table of logarithms to the base 10. Such logarithms are those which are usually adopted for the purposes of practical calculation.

Natural logarithms were first devised by John Napier, who published his invention in 1614.[1] Loga-

---

[1] *Logarithmorum Canonis Descriptio.* 4to. Edinburgh, 1614.

rithms to the base 10 are now used in all but the simplest numerical calculations which it is needful to make in the exact sciences; their value arises solely from the fact that addition and subtraction are easier operations than multiplication and division.

§ 11. *The Cartesian Method of Determining Position.*

(γ) In order to determine the position of a point P_1 in space of two dimensions, we may draw the line B A B′, joining the given points A B and another line C A C′ at right angles to this through A. These will divide the plane into four equal portions termed *quadrants*. Let P_1 M be a line drawn from the point P_1 (the position of

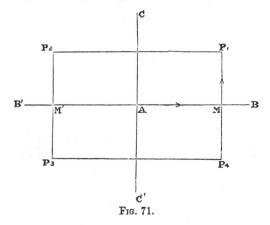

Fig. 71.

which relative to A we wish to determine), parallel to C A and meeting B′A B in M. Then we may state the following rule to get from A to P_1: Take a step A M from A on the line B′A B, and then a step to the left at right angles to this equal to M P_1. Now a step like A M may be taken either forwards along A B or backwards along A B′. Precisely as before (see p. 100) we shall take $+\mathrm{A M}$ to mean a step *forwards* along $\mathrm{A B}$, and $-\mathrm{A M}$ to mean a step $\mathrm{A M'}$ *backwards* along $\mathrm{A B'}$ through the same distance $\mathrm{A M}$. Let us use the letter i to denote the operation, which we have represented by $(\pi/2)$ on p. 151. Thus applied to unit step it will signify : Step *forwards* in the direction of the previous step and from its finish unit distance, and then rotate this unit distance through a right angle counter-clockwise about the finish of the previous step. The operator i placed before a step, thus $i \cdot \mathrm{M P_1}$, will then be interpreted as follows : Step from M in the direction $\mathrm{A B}$ a distance equal to the length $\mathrm{M P_1}$, and then rotate this step $\mathrm{M P_1}$ about M counter-clockwise through a right angle. We are thus able to express symbolically the position of $\mathrm{P_1}$ relative to A, or the step $\mathrm{A P_1}$, by the relation

$$\mathrm{A P_1} = \mathrm{A M} + i \cdot \mathrm{M P_1}.$$

If we had to get to a point $\mathrm{P_4}$ in the quadrant $\mathrm{B A C'}$, instead of to $\mathrm{P_1}$, we should have, instead of stepping forwards from M, to step *backwards* a distance $\mathrm{M P_4}$, and then rotate this through a right angle counter-clockwise. The step backwards would be denoted by inserting a $-$ sign as a reversing operation (see p. 39), and we should have

$$\mathrm{A P_4} = \mathrm{A M} - i \cdot \mathrm{M P_4}.$$

Next let us see how we should get to a point like $\mathrm{P_2}$ in the quadrant $\mathrm{C A B'}$, where $\mathrm{P_2}$ is at a perpendicular distance $\mathrm{P_2 M'}$ from $\mathrm{A B'}$. First, we must take a step, $\mathrm{A M'}$, backwards ; this is denoted by $-\mathrm{A M'}$; secondly, we must step *forwards* from $\mathrm{M'}$ a distance $\mathrm{M' P_2}$; since this step is *forwards*, it will be towards A ; thirdly, by applying the operation i to this step, we rotate it about

M' counter-clockwise through a right angle, and so reach P_2. Hence

$$A P_2 = - A M' + i \cdot M' P_2.$$

Finally, if we wish to reach P_3 in the quadrant B'A C', we must step backwards A M', and then still further backwards a step M' P_3, and lastly rotate this step counter-clockwise through a right angle. This will be expressed by

$$A P_3 = - A M' - i \cdot M' P_3.$$

Now let us suppose P_1, P_2, P_3, P_4, to be the four corners of a rectangular figure whose centre is at A and whose sides are parallel to B A B' and C A C'. Let the number of units in A M be x, and the number in M P_1 be y, then we may represent the four steps which determine the positions of the P's relative to A as follows :—

$$A P_1 = x + i y \qquad A P_2 = - x + i y$$
$$A P_3 = - x - i y \qquad A P_4 = x - i y.$$

Here x and y are mere numbers, but, when we represent these numbers by steps on a line, the y-numbers are to be taken on a certain line at right angles to that line on which the x-numbers are taken. Thus the moment we represent our x and y numbers by lengths, they give us a means of determining position.

The quantities x and y might thus be used to determine the position of a point, if we supposed them to carry with them proper signs. Our general rule would then be to step forwards from A along A B a distance x, and then from the end of x a distance forwards equal to y; rotate this step y about the end of x counter-clockwise through a right angle, and the finish of y will then be the point determined by the quantities x, y.

If x or y be negative, the corresponding forwards must be read : Step forwards a negative quantity, that is, step backwards. Thus :—

P₁, or position in the quadrant B A C is determined by x, y.

P₂	C A B′	$-x, y$.
P₃	B′ A C′	$-x, -y$.
P₄	C′ A B	$x, -y$.

The quantities x and y are termed the *Cartesian co-ordinates* of the point P, this method of determining the

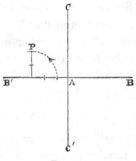

FIG. 72.

position of a point having been first used by Descartes. B A B′ and C A C′ are termed the co-ordinate *axes* of x and y respectively, while A is called the *origin* of co-ordinates. For example, let the Cartesian co-ordinates of a point be $(-3, 2)$. How shall we get at it from the origin A? If P be the point, we have A P $= -3 + i \cdot 2$. Hence we must step backwards 3 units; from this point step forwards 2 and rotate this step 2 about the extremity of the step 3 through a right angle counterclockwise; we shall then be at the required point.

If P be determined by its Cartesian co-ordinates x and y, we might find a succession of points, P, by always taking a step y related in a certain invariable fashion to any step x which has been previously made.

Such a succession of points P, obtained by giving x every possible value, will form a line or curve, and the relation between x and y is termed its *Cartesian equation*.

As an instance of this, suppose that for every step x, we take a step y equal to the double of it. Then we shall have for our relation $y = 2\,x$, and our instructions

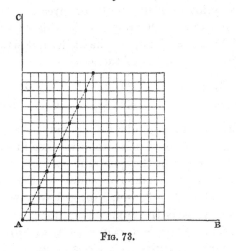

Fig. 73.

to reach any point P of the series are $x + i \cdot 2\,x$. Suppose the quadrant B A C divided into a number of little squares by lines parallel to the axes, and let us take the sides of these squares to be of unit length. Then if we take in succession $x = 1$, 2, 3, &c., we can easily mark off our steps. Thus : 1 along A B and then 2 to the left; 2 along A B and 4 to the left; 3 along A B and then 6 to the left; 4 along A B and then 8 to the left; 5 along A B and then 10 to the left, and so on. It will be obvious (by p. 106) that our points all lie upon a straight line through A, and however many steps we take along A B, followed by double steps perpendicular to it, we shall always arrive at a point on the same line. If we gave x negative values we should obtain that part of the line which lies in the third quadrant B'A C'. Hence we see that $y = 2x$ is the equation to a straight line which passes through A.

Let us take another example. Suppose that the rectangle contained by y and a length of 2 units, always contains as many units of area as there are square units in x^2. Our relation in this case may be expressed by $2y = x^2$, and we have the following series of steps from A to points of the series :—

$$1+i.\tfrac{1}{2}, \qquad 2+i.2, \qquad 3+i.\tfrac{9}{2},$$
$$4+i.8, \qquad 5+i.\tfrac{25}{2}, \qquad 6+i.18, \&c.$$

We can by means of our little squares easily follow out the operations above indicated ; we thus find a series of points like those in the quadrant B A C of the figure. If however we had taken x equal to the negative quantities $-1, -2, -3, -4, -5, -6$, &c., we should have found precisely the same values for y, because we have seen that $(-a) \times (-a) = a^2 = (+a) \times (+a)$. These negative values for x give us a series of points like those in the quadrant B'A C of the figure. It is impossible that any points of the series should lie below B A B', because both negative and positive values for x give when squared a positive value for the step y, so that no possible x-step would give a negative y-step. The series of points obtained in this fashion are found to lie upon a curve which is one of those shadows of a circle which we have termed parabolas.

Hence we may say that $2y = x^2$ is the equation to a parabola.

This method of plotting out curves is of great value, and is largely used in many branches of physical investigation. For example, if the differences of successive x-steps denote successive intervals of time, and y-steps the corresponding heights of the column of mercury in a barometer above some chosen mean position,

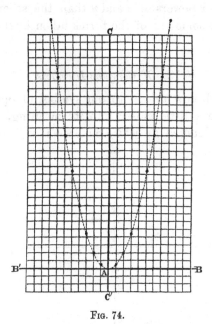

Fig. 74.

the series of points obtained will, if the intervals of time be taken small enough, present the appearance of a curve. This curve gives a graphical representation of the variations of the barometer for the whole period during which its heights have been plotted out. Barometric curves for the preceding day are now given in several of the morning papers. Heights corresponding to each instant of time are in this case generally registered automatically by means of a simple photographic apparatus.

The plotting out of curves from their Cartesian equations, usually termed *curve tracing*, forms an extremely interesting portion of pure mathematics. It may be shown that any relation, which does not involve higher powers of *x* and *y* than the second, is the equation to some one of the forms taken by the shadow of a circle.

§ 12. *Of Complex Numbers.*

We shall now return to our symbol of operation *i*, and inquire a little closer into its meaning. Let the point P be denoted as before by A M + *i* . M P, so that we

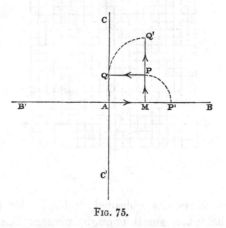

Fig. 75.

should read this result : Step from A to M along A B, and from M to P′ along the same line (where M P′ = M P), finally rotate M P′ about M counter-clockwise through a right angle ; M P′ will then take up the position M P. Now let M Q′ be taken equal to A P′, then A M + *i* . M Q′ will mean : Step from A to M and then from M perpendicular to A M to the left through a distance, M Q', equal to A P'. Since however M Q' = A P' = A M + M P = M P + P Q', P Q' must be equal to A M and we can read our operation

$$\text{A M} + i \,.\, (\text{M P} + \text{P Q}'),$$

which denotes two successive steps at right angles to A M, namely M P followed by the step P Q'. Suppose now we wished to rotate this latter step through a right angle counter-clockwise, we should have to introduce before it the symbol i, and M P + i . P Q' would signify the step M P followed by the step P Q at right angles to it to the left. Now P Q' is equal to A M, and hence the result of this operation must bring us to Q, a point on A C which might have been reached by the simple operation 0 + i . A Q. Thus we may put

$$0 + i \,.\, \text{A Q} = \text{A M} + i \,.\, (\text{M P} + i \,.\, \text{P Q})$$
$$= \text{A M} + i \,.\, \text{M P} + i \,.\, i \,.\, \text{P Q};$$

or, since the quantities A Q, A M, M P, and P Q here merely denote numerical magnitudes, and since as such A Q = M P and A M = P Q, we must have

$$0 = \text{A M} + i \,.\, i \,.\, \text{A M},$$
or
$$- \text{A M} = i \,.\, i \,.\, \text{A M}.$$

Thus the operation i is of such a character that repeated twice it is equivalent to a mere reversor, or, as we may express it symbolically,

$$- 1 = i^2.$$

This may be read in words: Turn a step counter-clockwise through a right angle, and then again counter-clockwise through another right angle, and we have the same result as if we had reversed the step. Now we have seen (p. 144) that if x be such a quantity that multiplied by itself it equals a, x is termed the square root of a, and written $\sqrt{a}$. Hence since $i^2 = -1$, we may write $i = \sqrt{-1}$.

This symbol is completely unintelligible so far as *quantity* is concerned; it can represent no quantity conceivable, for the squares of all conceivable quantities are positive quantities. For this reason $\sqrt{-1}$ is sometimes termed an *imaginary quantity*. Treated however as a *symbol of operation* $\sqrt{-1}$ has a perfectly clear and real meaning; it is here an instruction to step forwards a unit length and then rotate this length counter-clockwise through a right angle.

Any expression of the form $x + \sqrt{-1}\,y$ is termed a *complex number.*

Let P be any point determined by the step $A P = A M + \sqrt{-1}\,M P$, and let r, x, y be the numerical values of the lengths A P, A M, and P M. It follows from the right-angled triangle P A M that $r^2 = x^2 + y^2$. The quantity r is then termed the *modulus* of the complex number $x + \sqrt{-1}\,y$.

Further let the angle M A P contain θ units of angle; then

$$\sin\theta = \frac{P M}{A P} = \frac{y}{r}, \quad \cos\theta = \frac{A M}{A P} = \frac{x}{r},$$

or $y = r\sin\theta, \quad x = r\cos\theta.$

The angle θ is termed the *argument* of the complex number. Here r and θ are the polar co-ordinates of P, and we are thus able to connect them with the Cartesian co-ordinates; they are respectively the modulus and argument of the complex number which may be formed from the Cartesian co-ordinates. Since r is merely numerical we may write the complex number $x + \sqrt{-1}\,y$ in the form $r\,.\,(\cos\theta + \sqrt{-1}\sin\theta)$, or as the product of its modulus and the operator

$$\cos\theta + \sqrt{-1}\sin\theta,$$

which depends solely on its argument θ. Hence we may interpret the step

$$\text{A P} = r\,.\,(\cos\theta + \sqrt{-1}\sin\theta)$$

as obtained in the following fashion: Rotate unit length from A B through an angle θ, and then stretch it in the ratio of $r : 1$. The latter part of this operation

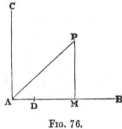

Fig. 76.

will be signified by the modulus r, the former by the operator $(\cos\theta + \sqrt{-1}\sin\theta)$. Thus if A D be of unit length and lying in A B, we may read—

$$\text{A P} = r\,.\,(\cos\theta + \sqrt{-1}\sin\theta)\,.\,\text{A D,}$$

and we look upon our complex number as a symbol denoting the combination of two operations performed on a unit step A D.

Starting then from the idea of a complex number as denoting position, we have been led to a new operation represented by the symbol $\cos\theta + \sqrt{-1}\sin\theta$. This is obviously a generalised form of our old symbol $\sqrt{-1}$. The operator $\cos\theta + \sqrt{-1}\sin\theta$ applied to any step bids us turn the step through an angle θ. We shall see that this new conception has important results.

§ 13. *On the Operation which turns a Step through a given Angle.*

Suppose we apply the operator $(\cos\theta + \sqrt{-1}\,\sin\theta)$ twice to a unit step. Then the symbolic expression for this operation will be

$$(\cos\theta + \sqrt{-1}\,\sin\theta)\,(\cos\theta + \sqrt{-1}\,\sin\theta),$$

or $\qquad\qquad (\cos\theta + \sqrt{-1}\,\sin\theta)^2.$

But to turn a step first through an angle θ and then through another angle θ is clearly the same operation as turning it by one rotation through an angle 2θ, or as applying the operator $\cos2\theta + \sqrt{-1}\,\sin2\theta$. Hence we are able to assert the equivalence of the operations expressed by the equation—

$$(\cos\theta + \sqrt{-1}\,\sin\theta)^2 = \cos2\theta + \sqrt{-1}\,\sin2\theta.$$

In like manner the result of turning a step by n operations through successive angles equal to θ must be identical with the result of turning it at once through an angle equal to n times θ, or we may write

$$(\cos\theta + \sqrt{-1}\,\sin\theta)^n = \cos n\theta + \sqrt{-1}\,\sin n\theta.$$

This important equivalence of operations was first expressed in the above symbolical form by De Moivre, and it is usually called after him De Moivre's Theorem.

We are now able to consider the operation by means of which a step A P can be transformed into another A Q. We must obviously turn A P about A counter-clockwise till it coincides in position with A Q; in this case P will fall on P′, so that A P′ = A P. Then we must stretch A P′ into A Q; this will be a process of multiplying it by some quantity ρ, which is equal to the ratio of A Q to A P′.

Expressing this symbolically, if ϕ be the angle P A Q, we have

$$(\cos\phi + \sqrt{-1}\sin\phi) \, . \, \text{A P} = \text{A P}'.$$
$$\rho \, . \, (\cos\phi + \sqrt{-1}\sin\phi) \, . \, \text{A P} = \rho \, . \, \text{A P}' = \text{A Q}.$$

This last equation we can interpret in various ways:

(i) $\rho \, . \, (\cos\phi + \sqrt{-1}\sin\phi)$ is a complex number of which ρ is the modulus and ϕ the argument. Hence we may say that to multiply a step by a complex number is to turn the step through an angle equal to the argument and to alter its length by a stretch represented by the modulus.

(ii) Or, again, we may consider the step A P as itself representing a complex number, $x + \sqrt{-1}\,y$, or if r be

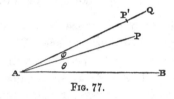

Fig. 77.

the scalar value of A P and θ the angle B A P, we may put $\text{A P} = r(\cos\theta + \sqrt{-1}\sin\theta)$. Similarly A Q will be a complex number, and its scalar magnitude ($= \rho \, . \, \text{A P}' = \rho \, r$) will be its modulus, while the angle B A Q $= \theta + \phi$ will be its argument. We have then the following identity—

$$\rho\,(\cos\phi + \sqrt{-1}\sin\phi) \, . \, r\,(\cos\phi + \sqrt{-1}\sin\theta) =$$
$$\rho\,r \, . \, (\cos\overline{\theta + \phi} + \sqrt{-1}\,\cos\overline{\phi + \theta}).$$

This may be read in two ways:

First, the product of two complex numbers is itself a complex number, and has the product of the moduli for its modulus, the sum of the arguments for its argument.

O

Or secondly, if we turn unit step through an angle θ and give a stretch r, and then turn the result obtained through an angle ϕ and give it a stretch ρ, the result will be the same as turning unit step through an angle $\theta + \phi$ and giving it a stretch equal to $\rho\, r$.

Thus we see that any relation between complex numbers may be treated either as an algebraical fact relating to such numbers, or as a theorem concerning operations of turning and stretching unit steps.

(iii) We may consider what answer the above identity gives to the question: What is the ratio of two *directed* steps A Q and A P? Or, using the notation suggested on p. 45, we ask: What is the meaning of the symbol $\dfrac{\text{AQ} \mid}{\mid \text{AP}}$? A step like A P (or A Q) which has magnitude, direction, and sense is, as we have noted, termed a vector. We therefore ask: What is the ratio of two vectors, or what operation will convert one into the other? The answer is: An operation which is the product of a turning (or spin) and a stretch. Now the stretch is a scalar quantity, a numerical ratio by which the scalar magnitude of A P is connected with that of A Q. The stretch therefore is a scalar operation. Further, the turning or spin converts the direction of A P into that of A Q, and it obviously takes place by spinning A P round an axis perpendicular to the plane of the paper in which both A P and A Q lie. Thus the second part of the operation by which we convert A P into A Q denotes a spin (counter-clockwise) through a definite angle about a certain axis. The amount of the spin might be measured by a step taken along that axis. Thus, for instance, if the spin were through 6 units of angle, we might measure 6 units of length along the axis to

denote its amount. We may also agree to take this
length along one direction of the axis ('out from the
face of the clock') if the spin be counter-clockwise, and
in the opposite direction ('behind the face of the clock')
if the spin be clockwise. Thus we see that our spinning
operation may be denoted by a line or step having
magnitude, direction, and sense; that is, by a *vector*.
We are now able to understand the nature of the ratio
of two vectors; it is an operation consisting of the product of a scalar and a vector. This product was termed
by Sir William Hamilton a *quaternion*, and made the
foundation of a very powerful calculus.

Thus a quaternion is primarily the operation which
converts one vector step into another. It does this by
means of a spin and a stretch. If we have three points
in plane space, the reader will now understand how
the position of the third with regard to the first can be
made identical with that of the second by means of a
spin and a stretch of the step joining the first to the
third, that is, by means of a quaternion.[1]

§ 14. *Relation of the Spin to the Logarithmic Growth of Unit Step.*

Let us take a circle of unit radius and endeavour
to find how its radius grows in describing unit angle
about the centre. Hitherto we have treated of growth
only in the direction of length; and hence it might be
supposed that the radius of a circle does not 'grow' at
all as it revolves about the centre. But our method of
adding vector steps suggests at once an obvious extension
of our conception of growth. Let a step A P become

[1] The term 'stretch' must be considered to include a squeeze or a
stretch denoted by a scalar quantity ρ less than unity.

A Q as it rotates about A through the angle P A Q, then if we marked off A Q a distance A P' equal to A P, P' Q would be the *scalar* growth of A P; that is, its growth

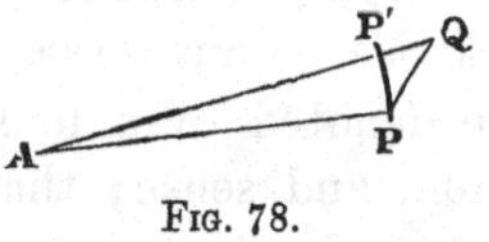

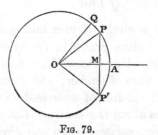

FIG. 78.

in the direction of its length.   But if A P be treated as a vector (see p. 153)

$$A Q = A P + P Q,$$

or the *directed* step P Q must be added to A P in order to convert it into A Q; P Q may be thus termed the directed growth of A P.   If we join P P', we shall have P Q equal to the sum of P P' and P' Q.   Now if the angle P A P' be taken very small P P' will be ultimately perpendicular to A P, and this part of the growth P Q might be represented by $\sqrt{-1} \cdot$ P P'.   Hence we are led to represent a growth perpendicular to a rotating line by a scalar quantity multiplied by the symbol $\sqrt{-1}$.

We can now consider the case of our circle of unit radius.   Let O P be a radius which has revolved through

FIG. 79.

an angle θ from a fixed radius O A, and let O Q be an adjacent position of O P such that the angle Q O P is very small.   Then P Q will be a small arc sensibly coincident

with the straight line P Q, and the line P Q will be to all
intents and purposes at right angles to O P. Hence to
obtain O Q we must take a step P Q at right angles to
O P. This we represent by $\sqrt{-1}$ Q P. Since the radius
of the circle is unity the arc Q P, which equals the
radius multiplied by the angle Q O P (see p. 143), must
equal the numerical value of the angle Q O P. Or the
growth of O P is given by $\sqrt{-1} \times$ angle Q O P. Now
according to our definition of growing at logarithmic
rate (see p. 176), since O P is equally multiplied in describing equal angles about O, it must be growing at
logarithmic rate. What is this logarithmic rate for
unit angle ?

It must equal $\dfrac{P Q}{O P}$ divided by the ratio of the angle

Q O P to unit angle $= \dfrac{P Q}{O P \times \text{angle} \, Q O P} = \sqrt{-1}$ since O P

is unity. Thus O P is growing at logarithmic rate $\sqrt{-1}$
as it describes unit angle ; that is to say, the result of
turning O P through unit angle might be *symbolically*
expressed by $e^{\sqrt{-1}}$. Hence the result of turning O P
through an angle θ must be $e^{\sqrt{-1}\theta}$. We may then write

$$O P = O A . e^{\sqrt{-1}\theta}.$$

Drop P M perpendicular to O A and produce it to meet
the circle again in P′, then by symmetry M P = M P′, and
we have

$$O P = O M + \sqrt{-1} \, M P.$$
$$O P' = O M - \sqrt{-1} \, M P'.$$

Now since O P and O P′ are of unit magnitude,

$$\cos\theta = \frac{O M}{O P} = O M, \quad \sin\theta = \frac{P M}{O P} = P M.$$

Also the angle P′ O M equals the angle M O P, but, according to our convention as to the measurement of angles, it is of opposite sense, or equals $-\theta$. Thus we must write

$$O P' = O A \cdot e^{-\sqrt{-1}\theta}$$

Substituting their values, we deduce the symbolical results

$$\left.\begin{aligned} e^{\sqrt{-1}\theta} &= \cos\theta + \sqrt{-1}\ \sin\theta \\ e^{-\sqrt{-1}\theta} &= \cos\theta - \sqrt{-1}\ \sin\theta \end{aligned}\right\} \text{(i)}$$

Further,

$$O P - O P' = 2\sqrt{-1}\ P M$$
$$O P + O P' = 2\ O M ;$$

that is,

$$\left.\begin{aligned} e^{\sqrt{-1}\theta} - e^{-\sqrt{-1}\theta} &= 2\sqrt{-1}\ \sin\theta \\ e^{\sqrt{-1}\theta} + e^{-\sqrt{-1}\theta} &= 2\ \cos\theta \end{aligned}\right\} \text{(ii)}$$

These values for $\cos\theta$ and $\sin\theta$ in terms of the exponential e were first discovered by Euler. They are meaningless in the form (ii) when $\cos\theta$ and $\sin\theta$ are interpreted as mere numerical ratios; but they have a perfectly clear and definite meaning when we treat each side of the equation in form (i) as a symbol of operation. Thus $\cos\theta + \sqrt{-1}\ \sin\theta$ applied to unit step directs us to turn that step without altering its length through an angle θ; on the other hand, $e^{\sqrt{-1}\theta}$ applied to the same step causes it to grow at logarithmic rate unity *perpendicular* to itself, while it is turned through the angle θ. The two processes give the same result.

§ 15. *On the Multiplication of Vectors.*

We have discussed how vector steps are to be added, and proved that the order of addition is indifferent; we have also examined the operation denoted by the ratio of two vectors. The reader will naturally ask : Can no meaning be given to the product of two vectors ?

If both the vectors be treated as complex numbers, or as denoting operations, we have interpreted their product (see p. 193) as another complex number or as a resultant operation. Or again we have interpreted the product of two vectors when one denotes an operation and the other a step of position; the product in this case is a direction to spin the step through a certain angle and then stretch it in a certain ratio. But neither of these cases explains what we are to understand by the product of two steps of position.

Let A P, A Q be two such steps : What is the meaning of the product AP . AQ? Were A P and A Q merely

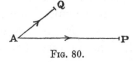

FIG. 80.

scalar quantities then their product would be purely scalar, and we should have no difficulty in interpreting the result A P . P Q as another scalar quantity. But when we consider the steps A P, P Q to possess not only

FIG. 81.

magnitude but *direction*, the meaning of their product is by no means so obvious.

If A Q were at right angles to A P (see fig. 81), we should naturally interpret the product A P . A Q as the

area of the rectangle on A Q and A P, or as the area of
the figure Q A P R. Now let us see how this area might
be generated. Were we to move the step A Q parallel
to itself and so that its end A always remained in the
step A P, it would describe the rectangle Q A P R while its
foot A described the step A P. Hence if A P and A Q are
at right angles we might interpret their product as
follows :

The product A P . A Q bids us move the step A Q
parallel to itself so that its end A traverses the step A P ;
the area traced out by A Q during this motion is the
value of the product A P . A Q.

It will be noted at once that this interpretation,
although suggested by the case of the angle Q A P being
a right angle, is entirely independent of what that angle
may be. If Q A P be not a right angle the area traced
out according to the above rule would be the parallelogram on A P, A Q as sides. Hence the interpretation we
have discovered for the product A P . A Q gives us an
intelligible meaning, whatever be the angle Q A P.

There is, however, a difficulty which we have not yet
solved. An area is a *directed* quantity (see p. 134), and
its direction depends on how we go round its perimeter.
Now the area Q A P R will be positive if we go round its
perimeter counter-clockwise, or from A to P ; that is, in

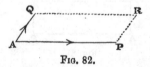

Fig. 82.

the direction of the first step of the product or in the
direction of motion of the second or moving step. Thus
the product A P . A Q will be the area Q A P R taken with
the sign suggested by the step A P. The product A Q . A P

will be formed by causing the step A P to move
parallel to itself along A Q, and it is therefore also the
area of the parallelogram on A Q and A P; but it is to be
taken with the sign suggested by A Q, or it is the area
P A Q R.

By our convention as to the sign of areas,

$$P A Q R = - Q A P R,$$
or $$A Q . A P = - A P . A Q.$$

Hence we see that, with the above interpretation, the
product of two vectors does not follow the commutative
law (see p. 45).

If we suppose the angle Q A P to vanish, and the
vector A Q to become identical with A P, the area of
the enclosed parallelogram will obviously vanish also.
Thus, if a vector step be multiplied by itself, the product
is zero; that is,

$$A P . A P = (A P)^2 = 0.$$

If we take a series of vector steps, a, β, γ, δ, &c.
then relations of the following types will hold among
them :

$$a^2 = 0, \qquad \beta^2 = 0, \qquad \gamma^2 = 0, \qquad \delta^2 = 0, \&c.$$
$$a \beta = - \beta a, \quad a \gamma = - \gamma a, \quad \beta \gamma = - \gamma \beta,$$
$$\delta \gamma = - \gamma \delta, \&c.$$

A series of quantities for which these relations hold
was first made use of by Grassmann, and termed by
him *alternate units*.

The reader will at once observe that alternate units
have an algebra of their own. They dispense with
the commutative law, or rather replace it by another
in which the sign of a product is made to alternate with
the alternation of its components. Their consideration
will suggest to the reader that the rules of arithmetic,

which he is perhaps accustomed to assume as necessarily true for all forms of symbolic quantity, have only
the comparatively small field of application to scalar
magnitudes. It becomes necessary to consider them as
mere conventions, or even to lay them aside entirely as
we proceed step by step to enlarge the meaning of the
symbols we are employing.

Although $2 \times 2 = 0$ and $2 \times 3 = -3 \times 2$ may be sheer
nonsense when 2 and 3 are treated as mere numbers, it
yet becomes downright common sense when 2 and 3 are
treated as directed steps in a plane.

Let us take two alternate units a, β and interpret
the quantity $a\, a + b\, \beta$, where a and b are merely scalar

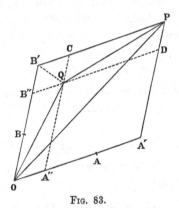

Fig. 83.

magnitudes. If o a be the vector a, $a\, a$ signifies that
we are to stretch o a to o a′ in the ratio of 1 to a. To
this o a′ we are to add the vector o b′ derived from o b
by giving it the stretch b. Hence if a′ p = o b′ the
vector o p represents the quantity $a\, a + b\, \beta$, which is
termed an *alternate number*. Let o q represent a second
alternate number $a′\, a + b′\, \beta$, obtained by adding the
results of applying two other stretches $a′$ and $b′$ to the

alternate units a and β. In the same way we might obtain, by adding the results of stretching three alternate units (a, β, γ), alternate numbers with three terms (of the form $a\,a + b\,\beta + c\,\gamma$), and so on. If we take the *product* of as many alternate numbers as we have used alternate units in their composition, we obtain a quantity called a *determinant*, which plays a great part in the modern theory of quantity. We shall confine ourselves here to the consideration of a determinant formed from two alternate units. Such a determinant will be represented by the product O P . O Q, which according to our convention as to the multiplication of vectors equals the area of the parallelogram on O P, O Q as sides, or (by p. 122) twice the triangle Q O P. Through Q draw C Q A″ parallel to O B, and D Q B″ parallel to O A, then O A″ = $a'\,a$ and O B″ = $b'\,\beta$. Join B′ Q, then twice the triangle B′ Q P equals the parallelogram B″ P. Hence, adding to both these the parallelogram A′ B″ we have the parallelogram A′ B″ together with twice the triangle B′ Q P equal to the parallelogram B′ A′, or to twice the triangle B′ O P. But the triangle B′ O P equals the sum of the triangles O Q B′, B′ Q P, and O P Q. It follows then that the parallelogram A′ B″ must equal twice the triangle O P Q together with twice the triangle O Q B′. Now twice the latter equals B′ A″. Hence the difference of the parallelograms A′ B″ and B′ A″ is equal to twice O P Q. The parallelogram A′ B″ is obtained from the parallelogram A B by giving it two stretches a and b' parallel to its sides, and therefore its area equals $a\,b'$ times the area A B. Similarly B′ A″ equals $b\,a'$ times the area A B; but the area A B itself is $a\,\beta$. Thus we see that the identity

$$\text{O P . O' Q} = \text{A' B''} - \text{B' A''}$$

may be read

$$(a \, a + b \, \beta) \, (a' \, a + b' \, \beta) = (a \, b' - b \, a') \, a \, \beta.$$

Or, the determinant is equal to the parallelogram on the alternate units magnified in the ratio of 1 to $a \, b' - b \, a'$. It obviously vanishes if $a \, b' - b \, a' = 0$, or if $a/b = a'/b'$. In this case P and Q lie, by the property of similar triangles, on the same straight line through o, and therefore, as we should expect, the determinant o P . o Q is zero.

The reader will find little difficulty in discovering like properties for a determinant formed from three alternate units. In this case there will be a geometrical relation between certain volumes, which may be obtained by stretches in the manner explained on p. 139.[1]

We have in this section arrived at a legitimate interpretation of the product of two directed steps or vectors. We find that their product is an area, or according to our previous convention (see p. 134), also a directed step or vector whose direction is perpendicular to the plane which contains both steps of the product.

§ 16. *Another Interpretation of the Product of Two Vectors.*

The reader must remember, however, that the result of the preceding paragraph has only been obtained *by means of a convention* ; namely, by adopting the area of a certain parallelogram as the interpretation of the vector

---

[1] I have to thank my friend Mr. J. Rose-Innes for suggesting the introduction of the above remarks as to determinants. I may, perhaps, be allowed to add that by treating the alternate units, like Grassmann, as points, and the alternate number as their loaded centroid, a determinant of the second order is represented geometrically by a length, and we thus obtain for one of the *fourth* order a geometrical interpretation as a volume.

product. Only as long as we observe that convention
will our deductions with regard to the multiplication of
vectors be true. We might have adopted a different
convention, and should then have come to a different
result. It will be instructive to follow out the results
of adopting another convention, if only by so doing we
can impress the reader with the fact that the fundamental axioms of any branch of exact science are based
rather upon conventions than upon universal truths.

Suppose then that in interpreting the product
A P . A Q we consider A P to be a directed step which

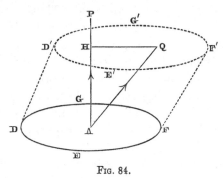

FIG. 84.

represents the area D E F G. This area will be perpendicular to the direction of A P, and we might assume as
our convention that the product A P . A Q shall mean the
volume traced out by the step A Q, moving parallel to
itself and in such wise that its end A takes up every
possible position in the plane D E F G. This volume will
be the portion of an oblique cylinder on the base D E F G
intercepted by a plane parallel to that base through Q.
We have seen (p. 141) that the volume of this cylinder
is the product of its base into its height, viz. the perpendicular distance A H between the two planes. Now
let r and ρ be the scalar magnitudes of A P and A Q

respectively, and θ = the angle P A Q. Then A H = $\rho \cos\theta$, and the volume = A P . A Q = $r \rho \cos\theta$, for r represents the number of units of area in D E F G. Hence, since a volume is a purely numerical quantity having only magnitude and no direction, we find that with this new convention the product of two vectors is a purely *scalar* quantity, or our new convention leads to a totally different result from the old.

Further, since r and ρ are merely numbers, $r \rho = \rho r$, and thus A P . A Q = $r \rho \cos\theta = \rho r \cos\theta$ = A Q . A P, if A Q be treated as the directed step which represents an area containing ρ units of area. Thus in this case the vector product obeys the commutative law, which again differs from our previous result. We can then treat the product of two vectors either as a vector and as a quantity not obeying the commutative law, or as a *scalar* and as a quantity obeying the commutative law. We are at liberty to adopt either convention, provided we maintain it consistently in our resulting investigations.

The method of varying our interpretation, which has been exemplified in the case of the product of two vectors, is peculiarly fruitful in the field of the exact sciences. Each new interpretation may lead us to vary our fundamental laws, and upon those varied fundamental laws we can build up a new calculus (algebraic or geometric as the case may be). The results of our new calculus will then be necessarily true for those quantities only for which we formulated our fundamental laws. Thus those laws which were formulated for pure number, and which, like the postulates of Euclid with regard to space, have been frequently supposed to be the only conceivable basis for a theory of quantity, are found to be true only within the limits of scalar magnitude. When we extend our conception of quantity and endow it with direction and position, we find those laws are no longer valid. We are compelled to suppose that one or more of those laws cease to hold or are replaced by others of a different form. In each case we vary the old form or adopt a new one to suit the wider interpretation we are giving to quantity or its symbols.

§ 17. *Position in Three-Dimensioned Space.*

Hitherto we have been considering only position in a plane; very little alteration will enable us to consider the position of a point P relative to a point A as determined by a step A P taken in space.

We may first remark, however, that while two points A and B are sufficient to determine in a plane the position of any third point P, we shall require, in order to fix the position of a point P in space, to be given three points A, B, C not lying in one straight line. If we knew only the distances of P from two points A and B, the point P might be anywhere on a certain circle which has its centre on the line A B and its plane perpendicular to that line; to determine the position of P on this circle, we require to know its distance from a third point C. Thus position in space requires us to have at least three non-collinear points (or such geometrical figures as are their equivalent) as basis for our determination of position. Space in which we live is termed space of three dimensions; it differs from space of two dimensions in requiring us to have three and not two points as a basis for determining position.

Three points will fix a plane, and hence if we are given three points A, B, C in space, the plane through them will be a definite plane separating all space into two halves. In one of these any point P whose position we require must lie. We may term one of these halves *below* the plane and the other *above* the plane. Let P N be the perpendicular from P upon the plane; then if we know how to find the point N in the plane A B C, the position of P will be fully determined so soon as we have settled whether the distance P N is to be measured above or below the plane. We may settle by convention that all distances above the plane shall be considered *positive*, and all below *negative*. Further, the position of the point N, upon which that of P depends, may be

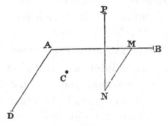

Fig. 85.

determined by any of the methods we have employed to fix position in a plane. Thus if N M be drawn perpendicular to A B, we have the following instruction to find the position of P: Take a step A M along A B, containing, say, x units; then take a step M N to the right and perpendicular to A B, but still in its plane, containing, say, y units; finally step *upwards* from N the distance N P perpendicular to the plane A B C, say, through z units. We shall then have reached the same point P as if we had taken the directed step A P. If x had been negative we should have had to step backwards from A; if y had been negative, perpendicular to A B only to the left; if z had been negative, perpendicular to the plane but *downwards.* The reader will easily convince himself that by observing these rules as to the sign of x, y, z he could get from A to any point in space.

Let i denote unit step along A B, j unit step to the right perpendicular to A B, but in the plane A B C, and k unit step perpendicular to the plane A B C upwards, from foot to head. Then we may write

$$A P = x \cdot i + y \cdot j + z \cdot k,$$

where x, y, z are scalar quantities possessing only magnitude and sign; but i, j, k are vector steps in three mutually rectangular directions.

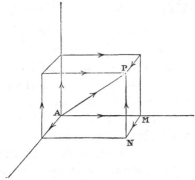

FIG. 86.

The step A P may be regarded as the diagonal of a solid rectangular figure (a *right six-face*, as we termed it on p. 138), and thus we shall get to the same point P by traversing any three of its non-parallel sides in succession starting from A. But this is equivalent to saying that the order in which we take the directed steps $x \cdot i$, $y \cdot j$, and $z \cdot k$ is indifferent.

The reader will readily recognise that the sum of a number of successive steps in space is the equivalent to the step which joins the start of the first to the

finish of the last; and thus a number of propositions
concerning steps in space similar to those we have
proved for steps in a plane may be deduced. By
dividing all space into little cubes by three systems of
planes mutually at right angles, we may plot out surfaces just as we plotted out curves. Thus we shall choose
any values we please for x and y, and suppose the
magnitude of the third step related in some constant
fashion to the previous steps. For example, if we take
the rectangle under z and some constant length a,
always equal to the differences of the squares on x and
y, or symbolically if we take $az = x^2 - y^2$, we shall
reach P by taking the step

$$\mathrm{A\,P} = x \cdot i + y \cdot j + \frac{x^2 - y^2}{a} \cdot k.$$

The series of points which we should obtain in this
way would be found to lie upon a surface resembling
the saddle-back we have described on p. 89. The
above relation between z, x, and y will then be termed
the *equation* to a saddle-back surface.

We cannot, however, enter fully on the theory of
steps in space without far exceeding the limits of our
present enterprise.

§ 18. *On Localised Vectors or Rotors.*

Hitherto we have considered the position of a point
P relative to a point A, and compared it with the
position of another point Q relative to the same point
A. Thus we have considered the ratio and product of
two steps A P and A Q.

We have thereby assumed either that the two steps
we were considering had a common extremity A, or at
least were capable of being moved parallel to themselves

till they had such a common extremity. Such steps are, as we have remarked, termed *vector* steps.

Suppose, however, that instead of comparing the position of two points P and Q relative to the same point A, we compared their positions relative to two different points A and B. The position of P relative to A will then be determined by the step A P and the position of Q relative to B by the step B Q.

Now it will be noted that these steps A P and B Q have not only direction and magnitude, but have themselves *position in space*. The step A P has itself position in space relative to the step B Q. It is no longer a step

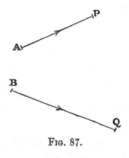

Fig. 87.

merely indicating the position of P with regard to A, but taken as a whole it has itself attained position when considered with regard to the step B Q. This *localising*, not of a point P relative to a point A, but of a step A P with regard to another step B Q, is a new and important conception. Such a localised vector is termed a *rotor* from the part it plays in the theory of rotating or spinning bodies.

Let us try and discover what operation will convert the rotor B Q into the rotor A P; in other words: What is the operation $\dfrac{A\ P}{B\ Q}$? In order to convert B Q into

A P we must make the magnitude and position of B Q the same as that of A P. Its magnitude may be made the same by means of a stretching operation which stretches B Q to A P. This stretch, as we have seen in the case of a quaternion (see p. 195), may be represented by a numerical ratio or a mere *scalar* quantity. Next let C D be the shortest distance between the rotors A P

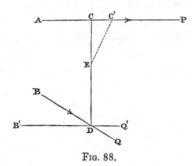

FIG. 88.

and B Q; then C D will be perpendicular to both of them.[1] B Q may then be made to coincide in position with A P by the following process:

First turn B Q about the shortest distance, C D, through some angle, Q D Q′, till it takes up the position B′Q′ parallel to A P; then slide B′Q′ along the

[1] That the shortest distance between two lines is perpendicular to both of them may be proved in the following manner. Let us suppose the lines replaced by perfectly smooth and very thin rods, and let two rings, one on either rod, be connected by a stretched elastic string. Obviously the rings will slide along the rods till the elastic string takes up the position of the shortest distance; for that will correspond to the least possible tension of the string. Suppose that the string is then not at right angles to one of the rods, say, at the point C. By holding the string firmly at E, we might shift the ring at C along the rod to C′, so that the angle E C′ C should be a right angle. Then since C′ is a right angle C E would be greater than C′ E, being the side opposite the greatest angle of the triangle E C′ C. Hence the length of string C′ E + E D is less than the length C D, or C D cannot be the shortest distance which we have supposed it to be. Thus the shortest distance must be at right angles to both lines.

shortest distance parallel to itself till its position coincides with A P. If we wished B′Q′ to coincide point for point with A P′, we should further have to slide it along A P till B′ and A were one.

Now the two operations of turning a line about another line at right angles to it, and moving it along that line, are just akin to the operations which are applied to the groove in the head of a screw when we drive the screw into a block of wood; or again to the handle of a corkscrew when we twist the screw into a cork. The handle in the one case and the groove in the other not only spin round, but go forward in the direction of the screw axis. Such a movement along an axis, and at the same time about it, is termed a *twist*. The ratio of the forward space described to the angle turned through during its description by the head of the screw is termed the *pitch* of the screw. This pitch will remain constant for all forward spaces described if the thread of the screw be uniform. Thus turn an ordinary corkscrew twice round, and it will have advanced twice as far through the cork as when it has been turned only once round. Let us see whether we cannot apply this conception of a screw to the operations by which we bring the rotor B Q into the position of the rotor A P. Upon a rod placed at C D, the shortest distance, suppose a fine screw cut with such a thread that its pitch equals the ratio of C D to the angle Q D Q′. Then if we suppose B Q attached to a nut upon this screw at D, when we turn B Q through the angle Q D Q′, the nut with B Q will advance (owing to the pitch we have chosen for the screw) through the distance D C. In other words, B Q will have been brought up to A P and coincide with it in position and direction.

Hence the operations by means of which B Q can be

made to coincide with A P are a stretch followed by a twist along a certain screw. A screw involves direction, position, and pitch; a twist (as of a nut) about this axis involves something additional, namely a magnitude, viz. that of the angle through which the nut is to be turned. Magnitude associated with a screw has been termed by the author of the present book a *motor*[1] (since it expresses the most general instantaneous motion of a rigid body). Hence the operation by which one rotor is converted into another may be described as a motor combined with a stretch. This operation stands in the same relation to two rotors as the quaternion to two vectors. The motor plays such an important part in several branches of physical inquiry that the reader will do well to familiarise himself with the conception.

The sum of two vector steps is, as we have seen (p. 153), a third vector; but unlike vector steps the sum of two rotors is in general a motor; only in special cases does it become either a rotor or a vector. The geometry of rotors and motors, which we have only here been able to hint at, forms the basis of the whole modern theory of the relative rest (Static) and the relative motion (Kinematic and Kinetic) of invariable systems.

§ 19. *On the Bending of Space.*

The peculiar topic of this chapter has been position, position namely of a point P relative to a point A. This relative position led naturally to a consideration of the geometry of steps. I proceeded on the hypothesis

---

[1] 'Preliminary Sketch of Biquaternions,' *Proceedings of the London Mathematical Society*, vol. iv. p. 383.

that all position is relative, and therefore to be determined only by a stepping process. The relativity of
position was a postulate deduced from the customary
methods of determining position, such methods in fact
always giving relative position. *Relativity of position
is thus a postulate derived from experience.* The late
Professor Clerk-Maxwell fully expressed the weight of
this postulate in the following words :—

All our knowledge, both of time and place, is relative.
When a man has acquired the habit of putting words together,
without troubling himself to form the thoughts which ought to
correspond to them, it is easy for him to frame an antithesis
between this relative knowledge and a so-called absolute knowledge, and to point out our ignorance of the absolute position of
a point as an instance of the limitation of our faculties. Any
one, however, who will try to imagine the state of a mind conscious of knowing the absolute position of a point will ever after
be content with our relative knowledge.[1]

It is of such great value to ascertain how far we can
be certain of the truth of our postulates in the exact
sciences that I shall ask the reader to return to our
conception of position albeit from a somewhat different
standpoint. I shall even ask him to attempt an exami-
nation of that state of mind which Professor Clerk-
Maxwell hinted at in his last sentence.

Suppose we had a tube of exceedingly small bore
bent into a circular shape, and within this tube a worm
of length A B. Then in the limiting case when we
make the bore of the tube and the worm infinitely fine,
we shall be considering space of one dimension. For
so soon as we have fixed *one* point, c, on the tube, the
length of arc c A suffices to determine the position of
the worm. Assuming that the worm is incapable of

Matter and Motion, p. 20.

recognising anything outside its own tube-space, it would still be able to draw certain inferences as to the nature of the space in which it existed were it capable of distinguishing some mark c on the side of its tube. Thus it would notice when it returned to the point c, and it would find that this return would continually recur as it went round in the bore; in other words, the worm would readily postulate the finiteness of space. Further, since the worm would always have the same *amount of bending*, since all parts of a circle are of the same shape, it might naturally assume the *sameness* of

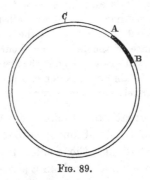

Fig. 89.

all space, or that space possessed the same properties at all points. This assumption is precisely akin to the one we make when we assert that the postulates of Euclidian geometry, which, experience teaches us, are practically true for the space immediately about us, are also true for all space; we assume the sameness of our three-dimensioned space. The worm would, however, have better reason for its postulate than we have, because it would have visited every part of its own one-dimensioned space.

Besides the finiteness and sameness of its space the worm might assert the relativity of position, and deter-

mine its position by the length of the arc between c
and A. Let us now make a variation in our problem
and suppose the worm incapable either of making or
of recognising any mark on the tube. Then it would
clearly be impossible for the worm to ascertain whether
its space were limited or not; it would never know
when it had made a complete revolution in its tube. In
fact, since the worm would always possess the same
amount of bending, it would naturally associate *that
bending with its physical constitution, and not with the
space which it was traversing.* It might thus very
reasonably suppose its space was infinite, or that it was
moving in an infinitely long tube. If the worm thus
associated bending with its physical condition it would
find no difference between motion in space of constant
bend (a circle) and motion in what is termed *homaloidal*
or flat space (a straight line); if suddenly transferred
from one to the other it would attribute the feeling
arising from difference of bending to some change
which had taken place in its physical constitution.
Hence in one-dimensioned space of constant bend all
position is necessarily relative, and the finite or infinite character of space will be postulated according as
it is possible or not to fix a point in it.[1]

Let us now suppose our worm moving in a different
sort of tube; for example, that shadow of a circle we
have called an ellipse. In such a tube the degree of
bending is not everywhere the same; the worm as it
passes from the place of least bending c to the place of
most bending D, will pass through a succession of bend-
ings, and each point H between c and D will have its

---

[1] This supposes the one-dimensioned space of constant bend to lie in a
plane; the argument does not apply to space like that of a *helix* (or the
form of a corkscrew), which is of constant bend, but yet not finite.

own degree of bending. Hence there is something
quite apart from the position of H relative to C which
characterises the point H ; namely, associated with H is
a particular degree of bending, and the position of the
point H in C D is at once fixed if we know the degree of
bending there. Thus the worm might determine *absolute* position in its space by the degree of bending associated with its position. The worm is now able to
appreciate differences of bend, and might even form a
scale of bending rising by equal differences. The zero
of such scale might be anywhere the worm pleased, and

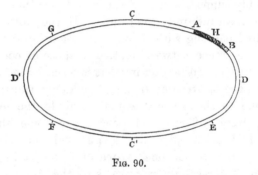

Fig. 90.

degrees of greater and less bend might be measured as
positive and negative quantities from that zero. This
zero might in fact be purely imaginary ; that is, represent
a degree of bending non-existent in the worm's space ;
for example, in the case of an ellipse, absolute straight-
ness, a conception which the worm might form as a
limit to its experience of degrees of bend.[1] Thus it
would seem that in space of 'varying bend,' or space
which is not same, position is not necessarily relative.
The relativity has ceased to belong to position in space ;
it has been transferred to the scale of bending formed

---

[1] Physicists may be reminded of the absolute zero of temperature.

by the worm; it has become a *relativity of physical feeling*. In the case of an elliptic tube there are owing to
its symmetry four points of equal bend, as H, E, F, and
G, but there is the following distinction between H, F
and E, G. If the worm be going round in the direction
indicated by the letters C H D E, at H or F it will be passing from positions of less to positions of greater bending,
but at E or G from positions of greater to positions
of less bending. Thus the worm might easily draw a
distinction between H, F and E, G. It would only be
liable to suppose the points H and F identical because

Fig. 91.

they possess the same degree of bending. We might
remove even this possible doubt by supposing the worm
to be moving in a pear-shaped tube, as in the accompanying figure; then there will only be two points of
equal bend, like H and G, which are readily distinguished
in the manner mentioned above.

We might thus conclude that in one-dimensioned
space of variable bend position is not necessarily
relative. There is, however, one point to be noted with
regard to this statement. We have assumed that the
worm will associate change of bending with change of
position in its space, but the worm would be sensible of

it as a change of physical state or as a change of feeling. Hence the worm might very readily be led into the error of postulating the sameness of its space, and attributing all the changes in its bend, really due to its position in space, to some periodic (if it moves uniformly round its tube) or irregular (if it moves in any fashion backwards and forwards) changes to which its physical constitution was subject. Similar results might also arise if the worm were either moving in space of the same bend, which bend could be changed by some external agency as a whole, or if again its space were of varying bend, which was also capable of changing in any fashion with time. The reader can picture these cases by supposing the tube made of flexible material. The worm might either attribute change in its degree of bend to change in the character of its space or to change in its physical condition not arising from its position in space. We conclude that the postulate of the relativity of position is not necessarily true for one-dimensioned space of varying bend.

When we proceed from one to two-dimensioned space, we obtain results of an exactly similar character. If we take perfectly even (so called *homaloidal*) space of two dimensions, that is, a plane, then a perfectly flat figure can be moved about anywhere in it without altering its shape. If by analogy to an infinitely thin worm we take an infinitely thin flat-fish, this fish would be incapable of determining position could it leave no landmarks in its plane space. So soon as it had fixed two points in its plane it would be able to determine *relative* position.

Now, suppose that instead of taking this homaloidal space of two dimensions we were still to take a perfectly same space but one of finite bend, that is, the surface of a sphere. Then let us so stretch and bend our flat-fish that it would fit on to some part of the sphere. Since the surface of the sphere is everywhere space of the same shape, the fish would then be capable of moving about on the surface without in any way alter-ing the amount of bending and stretching which we had found it necessary to apply to make the fish fit in any one position. Were the fish incapable of leaving landmarks on the surface of the sphere, it would be totally unable to determine position; if it could leave at least two landmarks it would be able to determine *relative* position. Just as the worm in the circular tube, the fish without landmarks might reasonably suppose its space infinite, or even look upon it as perfectly flat (homaloidal) and attribute the constant degree of bend and stretch to its physical nature.

Let us now pass to some space of two dimensions which is not same—to some space, for example, like the saddle-back surface we have considered on page 89, which has a varying bend. In this case the fish, if it fitted at one part of the surface, would not necessarily fit at another. If it moved about in its space, it would be needful that a continual process of bending and stretching should be carried on. Thus every part of this two-dimensioned space would be defined by the particular amount of bend and stretch necessary to make the fish fit it, or, as it is usually termed, by the *curvature*. In surfaces with some degree of symmetry there would necessarily be parts of equal curvature, and in some cases the fish might perhaps distinguish between these points in the same fashion as the worm distinguished between points of equal curvature in the case of an elliptic tube. In irregular surfaces, however, it is not necessary that such points of equal curvature should arise. We are thus led to conclusions like those we have formed for one-dimensioned space, namely : Position in space of two dimensions which is not same might be determined *absolutely* by means of the curvature. Our fish has only to carry about with it a scale of degrees of bending and stretching corresponding to various positions on the surface in order to determine absolutely its position in its space. On the other hand, the fish might very readily attribute all these changes of bend and stretch to variations of its physical nature in nowise dependent on its position in space. Thus it might believe itself to have a most varied physical life, a continual change of physical feeling quite independent of the geometrical character of the space in which it dwelt. It might suppose that space to be perfectly same, or even degrade it to the ' dreary infinity of a homaloid.' [1]

As a result, then, of our consideration of one and two-dimensioned space we find that, if these spaces be not same (*à fortiori* not homaloidal), we should by reason of their curvature have a means of determining absolute position. But we see also that a being existing in these dimensions would most probably attribute the effects of curvature to changes in its own physical condition in nowise connected with the geometrical character of its space.

What lesson may we learn by analogy for the three-dimensioned space in which we ourselves exist ? To begin with, we assume that all our space is perfectly *same*, or that solid figures do not change their shape in passing from one position in it to another. We base this postulate of sameness upon the results of observation

---

[1] In this case of two-dimensioned space assume it to be a plane. Cf. Clifford's *Lectures and Essays*, vol. i. p. 323.

in that somewhat limited portion of space of which we are cognisant.[1] Supposing our observations to be correct, it by no means follows that because the portion of space of which we are cognisant is for practical purposes same, that therefore *all* space is same.[2] Such an assumption is a mere dogmatic extension to the unknown of a postulate, which may perhaps be true for the space upon which we can experiment. To make such dogmatic assertions with regard to the unknown is rather characteristic of the mediæval theologian than of the modern scientist. On the like basis with this postulate as to the sameness of our space stands the further assumption that it is homaloidal. When we assert that our space is everywhere same, we suppose it of constant curvature (like the circle as one and the sphere as two-dimensioned space) ; when we suppose it homaloidal we assume that this curvature is zero (like the line as one and the plane as two-dimensioned space). This assumption appears in our geometry under the form that two parallel planes, or two parallel lines in the same plane—

[1] It may be held by some that the postulate of the sameness of our space is based upon the fact that no one has hitherto been able to form any geometrical conception of space-curvature. Apart from the fact that mankind habitually assumes many things of which it can form no geometrical conception (mathematicians the circular points at infinity, theologians transubstantiation), I may remark that we cannot expect any being to form a geometrical conception of the curvature of his space till he views it from space of a higher dimension, that is, practically, never.

[2] Yet it must be noted that, because a solid figure *appears* to us to retain the same shape when it is moved about in that portion of space with which we are acquainted, it does not follow that the figure *really* does retain its shape. The changes of shape may be either imperceptible for those distances through which we are able to move the figure, or if they do take place we may attribute them to 'physical causes'—to heat, light, or magnetism—which may possibly be mere names for variations in the curvature of our space.

that is, planes, or lines in the same plane, which however far produced will never meet—have a *real* existence
in our space. This real existence, of which it is clearly
impossible for us to be cognisant, we postulate as a
result built upon our experience of what happens in
a limited portion of space. We may postulate that
the portion of space of which we are cognisant is
practically homaloidal, but we have clearly no right
to dogmatically extend this postulate to *all* space. A
constant curvature, imperceptible for that portion of
space upon which we can experiment, or even a curvature which may vary in an almost imperceptible
manner with the time, would seem to satisfy all that
experience has taught us to be true of the space in
which we dwell.

But we may press our analogy a step further,
and ask, since our hypothetical worm and fish might
very readily attribute the effects of changes in the
bending of their spaces to changes in their own physical condition, whether we may not in like fashion be
treating merely as physical variations effects which are
really due to changes in the curvature of our space ;
whether, in fact, some or all of those causes which we
term physical may not be due to the geometrical construction of our space. There are three kinds of
variation in the curvature of our space which we ought
to consider as within the range of possibility.

(i) Our space is perhaps really possessed of a curvature varying from point to point, which we fail to appreciate because we are acquainted with only a small
portion of space, or because we disguise its small variations under changes in our physical condition which we
do not connect with our change of position. The mind
that could recognise this varying curvature might be

assumed to know the absolute position of a point. For such a mind the postulate of the relativity of position would cease to have a meaning. It does not seem so hard to conceive such a state of mind as the late Professor Clerk-Maxwell would have had us believe. It would be one capable of distinguishing those so-called physical changes which are really geometrical or due to a change of position in space.

(ii) Our space may be really same (of equal curvature), but its degree of curvature may change as a whole with the time. In this way our geometry based on the sameness of space would still hold good for all parts of space, but the change of curvature might produce in space a succession of apparent physical changes.

(iii) We may conceive our space to have everywhere a nearly uniform curvature, but that slight variations of the curvature may occur from point to point, and themselves vary with the time. These variations of the curvature with the time may produce effects which we not unnaturally attribute to physical causes independent of the geometry of our space. We might even go so far as to assign to this variation of the curvature of space 'what really happens in that phenomenon which we term the motion of matter.'[1]

[1] This remarkable *possibility* seems first to have been suggested by Professor Clifford in a paper presented to the Cambridge Philosophical Society in 1870 (*Mathematical Papers*, p. 21). I may add the following remarks: The most notable physical quantities which vary with position and time are heat, light, and electro-magnetism. It is these that we ought peculiarly to consider when seeking for any physical changes, which may be due to changes in the curvature of space. If we suppose the boundary of any arbitrary figure in space to be distorted by the variation of space-curvature, there would, by analogy from one and two dimensions, be no change in the volume of the figure arising from such distortion. Further, if we *assume* as an axiom that space resists curvature with a resistance

We have introduced these considerations as to the nature of our space to bring home to the reader the character of the postulates we make in the exact sciences. These postulates are *not*, as too often assumed, necessary and universal truths; they are merely axioms based on our experience of a certain limited region. Just as in any branch of physical inquiry we start by making experiments, and basing on our experiments a set of axioms which form the foundation of an exact science, so in geometry our axioms are really, although less obviously, the result of experience. On this ground geometry has been properly termed at the commencement of Chapter II. a *physical* science. The danger of asserting dogmatically that an axiom based on the experience of a limited region holds universally will now be to some extent apparent to the reader. It may lead us to entirely overlook, or when suggested at once reject, a possible explanation of' phenomena. The hypotheses that space is not homaloidal, and again, that its geometrical character may change with the time, may or may not be destined to play a great part in the physics of the future; yet we cannot refuse to consider them as possible explanations of physical phenomena, because they may be opposed to the popular dogmatic belief in the universality of certain geometrical axioms—a belief which has arisen from centuries of indiscriminating worship of the genius of Euclid.

proportional to the change, we find that waves of 'space-displacement' are precisely similar to those of the elastic medium which we suppose to propagate light and heat. We also find that space-twist' is a quantity exactly corresponding to magnetic induction, and satisfying relations similar to those which hold for the magnetic field. It is a question whether physicists might not find it simpler to assume that space is capable of a varying curvature, and of a resistance to that variation, than to suppose the existence of a subtle medium pervading an invariable homaloidal space.

CHAPTER V.

MOTION.

§ 1. *On the Various Kinds of Motion.*

WHILE the chapters on Space and Position considered
the sizes, the shapes, and the distances of things, the
present chapter on Motion will treat of the changes in
these sizes, shapes, and distances, which take place from
time to time.

The difference between the ordinary meaning attached to the word 'change' in everyday life and the
meaning it has in the exact sciences is perhaps better
illustrated by the subject of this chapter than by any
other that we have yet studied. We attained exactness
in the description of quantity and position by substituting the method of representing them by straight lines
drawn on paper for the method of representing them by
means of numbers; though this, at first sight, might
easily seem to be a step backwards rather than a step
forwards, since it is more like a child's sign of opening
its arms to show that its stick is so long, than a process
of scientific calculation.

It is, however, by no means an easy thing to give
an accurate description of motion, even although it is
itself as common and familiar a conception as quantity
or position.

Let us take a simple case. Suppose that a man, on a
railway journey, is sitting at one end of a compartment

Q 2

with his face towards the engine; and that, while the
train is going along, he gets up and goes to the other
end of the compartment and sits down with his back to
the engine.   For ordinary purposes this description is
amply sufficient, but it is very far indeed from being
an exact description of the motion of the man during
that time.   In the first place, the train was moving,
and it is necessary to state in what direction, and how
fast it was going at every instant during the interval
considered.   Next, we must describe the motion of the
man relatively to the train; and, for this purpose, we
must neglect the motion of the train and consider how
the man would have moved if the train had been at
rest.   First of all, he changes his position from one
corner of the compartment to the opposite corner;
next, in doing this he turns round; and, lastly, as he is
walking along or rising up or sitting down, the size and
shape of many of his muscles are altered.   We should
thus have to say, first, exactly how fast and in what
direction he was moving at every instant, as we had to
do in the case of the train; then, how quickly he was
turning round; and, lastly, what changes of size or
shape were taking place in his muscles, and how fast
they were occurring.

It may be urged that this would be a very troublesome operation, and that nobody wants to describe the
motion of the man so exactly.   This is quite true; the
case which has been taken for illustration is not one
which it is necessary to describe exactly, but we can
easily find another case which is very analogous
to this, and which it is most important to describe
exactly.   The earth moves round the sun once in every
year; it is also rotating on its own axis once every day;
the floating parts of it—the ocean and the air—are

constantly undergoing changes of shape and state
which we can observe and which it is of the utmost
importance that we should be able to predict and
calculate; even the solid nucleus of the earth is constantly subject to slight changes in size and shape,
which, however, are not large enough to admit of accurate observation. Here, then, is a problem whose
complexity is quite as great as that of the former, and
whose solution is of pressing practical importance.

The method which is adopted for attacking this
problem of the accurate description of motion is to begin
with the simplest cases. By the simplest cases we mean
those in which certain complicating circumstances do
not arise. We may first of all restrict ourselves to the
study of the motions of those bodies in which there is
no change of size or shape. A body which preserves
its size and shape unaltered during the interval of time
considered is called a *rigid* body. The word 'rigid' is
here used in a technical sense belonging to the science
of dynamic, and does not mean, as in ordinary language, a body which resists alteration of size and shape,
but merely a body which, during a certain time,
happens not to be altered in those respects. Then, as the
first and simplest case, we should study that motion of
a rigid body in which there is no turning round, and
in which therefore every line in the body keeps the
same direction (though of course not the same position)
throughout the motion. We state this by saying that
every line 'rigidly connected' with the body remains
parallel to itself. Such a motion is called a *motion of
translation*, or simply a *translation*; and so the first and
simplest case we have to study is the translation of
rigid bodies. After that we must proceed to consider
their turning round, or *rotation*; and then we have to

describe the changes of size or shape which bodies may undergo, these last changes being called *strains*. The study of motion therefore requires the further study of translations, of rotations, and of strains, and further, the art of combining these together. When we have studied all this we shall be able to describe motions exactly ; and then, but not till then, will it be possible to state the exact circumstances under which motions of a given kind occur. The exact circumstances under which motions of a given kind occur we call a *law of nature*.

§ 2. *Translation and the Curve of Positions.*

Let us talk, to begin with, of the translation of a rigid body.

Suppose a table to be taken from the top to the bottom of a house in such a manner that the surface of it is always kept horizontal, and that its length is made always to point due north and south ; it may be taken down a staircase of any form, but it is not to be turned round or tilted up. The table will then undergo a translation. If we now consider a particular corner of the table, or the end of one of its legs, or any other point, this point will have described a certain curve in a certain manner ; that is to say, at every point of this curve it will have been going at a certain definite rate. Now the important property of a motion of translation, which makes it more easy to deal with than any other motion, is that for all points of the body this curve is the same in size and shape and mode of description. That this is so in the case of the table is at once seen from the fact that the table is never turned round nor tilted up during the motion, so that the different points of it must at any instant be moving in the same direction and at the same rate. In order therefore to describe this motion of the table it will be sufficient to describe the motion of any point of it, say the end of one of its legs. And so, in general, the problem of describing the motion of translation of any rigid body is reduced to the problem of describing the motion of a point along a curve.

Now this is a very much easier task than our original problem of describing the motion of the earth or the motion of the man in the train; but we shall see that, by properly studying this, it will be easy to build up out of it other more complicated cases. Still, even in this form our problem is not quite simple enough to be directly attacked. What we have to do, it must be remembered, is to state exactly where a certain point was, and how fast it was going at every instant of time during a certain interval. This would require us first to describe exactly the shape of the curve along which the point moved; next, to say how far it had travelled along the curve from the beginning up to any given instant; and lastly, how fast it was going at that instant. To deal with this problem we must first take the very simplest case of it, that, namely, in which the point moves along a straight line, and leave for the present out of account any description of the rate of motion of the point; so that we have only to say where the point was on a certain straight line at every instant of time within a given interval.

But we have already considered what is the best way of describing the position of a point upon a straight line. It is described by means of the step which is required to carry it to that position from a certain standard place, viz. a step from that place so far to the right or to the left. To specify the length of the step, if we are to describe it exactly, we must not make use of any words or numbers, but must draw a line which will represent the length corresponding to every instant of time within a certain interval, so that we may always be able to answer the question, Where was the point at this particular instant? But a question, in order to be exactly answered, must first be exactly asked; and to do this it is necessary that the instant of time about which the question is asked should be accurately specified.

Now time, like length, is a continuous quantity which cannot in general be described by words or numbers, but can be by the drawing of a line which shall represent it to a certain scale. Suppose, then, that the interval of time during which the motion of a point has to be described is the interval from twelve o'clock to one o'clock. We must mark on a straight line a point to represent twelve o'clock and another point to represent one o'clock; then every instant between twelve o'clock and one o'clock will be represented by a point which divides the distance between these two marked points in the same ratio in which that instant divides the interval between twelve o'clock and one o'clock. Then for every one of these points it is necessary to assign a certain length, representing (to some definite scale) the distance which the point has travelled up to that instant; and the question arises, In what way shall we mark down these lengths?

Let us first of all observe the difficulty of answering this question. If we could be content with an approximate solution instead of an exact one, we might make a table and put down in inches and decimals of an inch the distances travelled, making an entry for every minute, or even perhaps for every second during the

hour. Such tables are in fact constructed and published in the 'Nautical Almanac' for the positions of the
moon and of the planets. The labour of making this
table will evidently depend upon its degree of minute-
ness; it will of course take sixty times as long to make
a table showing the position of the point at every
second as to make one showing the position at every
minute, because there will be sixty times as many
values to calculate. But the problem of describing
exactly the motion of the point requires us to make a
table showing the position of the point at every instant;
that is, a table in which are entered an infinite number
of values. These values moreover are to be shown, not
in inches and decimals of an inch, but by lengths drawn
upon paper. Yet we shall find that this pictorial mode of
constructing the table is in most cases very much easier
than the other. We have only to decide where we shall
put the straight lines which represent the distances
that the point has travelled at different instants.

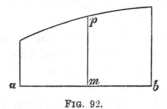

FIG. 92.

Let ab be the length which represents the interval
of time from twelve o'clock to one o'clock, and let m be
the point representing any intermediate instant. Then
if we draw at m a line perpendicular to ab whose length
shall represent (to any scale that we may choose) the
distance that the point has up to this instant travelled,
then p, the extremity of this line, will correspond to

an entry in our table. But if such lines be drawn
perpendicular to ab from every point in it, all the
points p, which are the several extremities of these
lines, will lie upon some curve; and this curve will represent an infinite number of entries in our table. For,
when once the curve is drawn, if a question is asked:
What was the position of the point at any instant
between twelve o'clock and one o'clock? (this instant
being specified in the right way by marking a point
between a and b which divides that line in the same
ratio as the given instant divides the hour), then the
answer to this question is obtained simply by drawing
a line through the marked point perpendicular to ab
until it meets the curve; and the length of that line
will represent, to the scale previously agreed upon, the
distance travelled by the point.

Such a curve is called the *curve of positions* for a
given motion of the point; and we arrive at this result,
that the proper way of specifying exactly a translation
along a straight line is to draw the curve of positions.

We have now learned to specify, by means of a
curve, the positions of a body which has motion of
translation along a straight line; and we have not
only represented an infinite number of positions instead of a finite number, which is all a numerical table
would admit, but have also represented each position
with absolute exactness instead of approximately. It
is important to notice that in this and in all similar
cases the exactness is ideal and not practical; it is
exactness of conception and not of actual measurement.
For though it is not possible to measure a given length
and to state that measure any more accurately by
drawing a line than it is by writing it down in inches
and decimals of an inch, yet the representation by

means of a line enables us to reason upon it with an exactness which would be impossible if we were restricted to numerical measurement.

§ 3. *Uniform Motion.*

Hitherto we have supposed our point to be moving along a straight line, but were it to move along a curve the construction given for the curve of positions would still hold good, only the distance traversed at any instant must now be measured from some standard position *along the curve.* Hence any motion of a point, or any motion of translation whatever, can be specified by a properly drawn curve of positions, and the problem of comparing and classifying different motions is therefore reduced to the problem of comparing and classifying curves. Here again it is advisable and even necessary to begin with a simple case. Let us take the case of uniform motion, in which the body passes over equal distances in equal times; and then, as we may easily see, the curve of positions is a straight line. Uniform motion may also be described as that in which a body always goes at the same rate, and not quicker at one time and slower at another. It is obvious that in this case any two equal distances would require equal times for traversing them, so that the two descriptions of uniform motion are equivalent.

It was shown by Archimedes (the proof is an easy one, depending upon the definition of the fourth proportional) that whenever equal distances are traversed in equal times, different distances will be traversed in times proportional to them. Assuming this proposition, it becomes clear that the curve of positions must be a straight line, for a straight line is the only curve which has the property that the height of every point of it is proportional to its horizontal distance from a fixed straight line.

We may also see in the following manner the connection between the straight line and uniform motion.

Suppose we walk up a hill so as always to get over a horizontal distance of four miles in an hour. The rate at which we go up will clearly depend on the steepness of the hill; and if the hill is a plane, *i.e.* is of the same steepness all the way up, then our rate of ascent will be the same at every instant, or our upward motion will be uniform. If the hill be four miles long and one mile high, then, since the four miles of horizontal distance will be traversed in an hour, the one mile of vertical distance will also be traversed in an hour, and we shall be gaining height at the uniform rate of one mile an hour. If the hill were two miles high, or, as we say twice as steep, then we should have been gaining height at the rate of two miles an hour. But now if we suppose a hill of varying steepness, so that the outline of it seen from one side is a curve, then it is clear that the rate at which we go up will depend upon the part of the hill where we are, assuming that the rate at which we go forward horizontally remains always the same. This ' elevation ' of the hill may be taken as the curve of positions for our vertical motion; for the horizontal distance that we have gone over, being always proportional to the time, may be taken to represent the time, and then the curve will have been constructed according to our rule, viz. a horizontal distance will have been taken proportional to the time elapsed, and from the end of this line a perpendicular will have been raised indicating the height which we have risen in that time. Uniform motion then has

for its curve of positions a straight line, and the rate
of the motion depends on the steepness of the line.
Variable motion, on the other hand, has a curved line
for its curve of positions, and the rate of motion
depends upon its varying steepness.

In the case of uniform motion it is very easy indeed
to understand what we mean by the rate of the motion.
Thus, if a man walks uniformly six miles an hour,
we know that he walks a mile in ten minutes, and the
tenth part of a mile in one minute, and so on in proportion. It may not, however, be possible to specify this
rate by means of numbers; that is to say, the man may
not walk any definite number of miles in the hour, and
the exact distance that he walks may not be capable of
representation in terms of miles and fractions of a mile.
In that case we shall have to represent the velocity or
rate at which the man walks in much the same way as
we have represented other continuous quantities. We
must draw to scale upon paper a line representing the
length that he has walked in an hour, or a minute, or
any other interval of time that we decide to select;
thus, for example, a uniform rate of walking might be
specified by marking points corresponding to particular
hours upon an Ordnance map. The rate of motion, or
velocity, is then a continuous quantity which can be
exactly specified, as we specify other continuous quantities, but which can be only approximately described by
means of numbers.

§ 4. *Variable Motion.*

Let us now suppose that the motion is not uniform,
and inquire what is meant in that case by the rate at
which a body moves.

A train, for example, starts from a station and in the course of a few minutes gets up to a speed of 30 miles an hour. It began by being at rest, and it ends by having this large velocity. What has happened to it in the meantime? We can understand already in a rough sort of way what is meant by saying that at a certain time between the two moments the train must have been going at 15 miles an hour, or at any other intermediate rate; but let us endeavour to make this conception a little more exact. Suppose, then, that a second train, which is indefinitely long, is moving in the same direction at a uniform rate of 15 miles an hour on a pair of rails parallel to that on which the first train moves; thus, when our first train is at rest the second one will appear to move past it at the rate of 15 miles an hour. When the first train starts an observer seated in it will see the second train going apparently rather more slowly than before, but it will still seem to be moving forwards. As the first train gets up its speed, this apparent forward motion will gradually decrease until the second train will appear to be going so slowly that conversation may be held between the two; this will take place when the rate of the first train has amounted to something nearly but not quite equal to 15 miles an hour, which we supposed to be the constant rate of the second train. But as the rate of the first train continues to increase there will come a certain instant at which the second train will appear to stop gaining upon the first and to begin to lose. At that particular instant it will be neither gaining nor losing, but will be going at the same rate; at that particular instant, therefore, we must say that the first train is going at the rate of 15 miles an hour. And it is at that instant only, for the equality of the rates does not last for any fraction of a second, however small; the very instant that the second train appears to stop gaining it also appears to begin losing. The two trains then run exactly together for no distance at all, not even for the smallest fraction of an inch, and yet we have to say that at one particular instant our first train is going at the rate of 15 miles an hour, although it does not continue to go at that rate during the smallest portion of time. There is no way of measuring this instantaneous velocity except that which has just been described of comparing the motion with a uniform motion having that particular velocity.

Upon this we have to make the very important remark that the rate at which a body is going is a property as purely instantaneous as is the precise position which it has at that instant. Thus, if a stone be let fall to the ground, at the moment that it hits the ground it is going at a certain definite rate; and yet at any previous moment it was not going so fast, since it does not move at that rate for the smallest fraction of a second. This consideration is somewhat difficult to grasp thoroughly, and in fact it has led many people to reject altogether the hypothesis of continuity; but still we may be helped somewhat in understanding it by means of our study of the curve of positions, wherein we saw that to a uniform motion corresponds a straight line and that the rate of the motion depends on the steepness of the line.

Let us now suppose a motion in which a body goes at a very slow but uniform rate for the first second, during the next second uniformly but somewhat faster, faster again during the third second, and so on. The curve of positions will then be represented by a series of straight lines becoming steeper and steeper and forming part of a polygon. From a sufficient distance off this polygon will look like a curved line ; and if, instead of taking intervals of a second during which the rates of motion are severally considered uniform, we had taken intervals of a tenth of a second, then the polygon would look like a curved line without our going so far away as before. For the shorter the lengths of the sides of our polygon, the more will it look curved, and if the intervals of time are reduced to one-tenth the sides will be only one-tenth as long. The rate at which the body under consideration is-moving when it is in the position to which any point of the polygon corresponds, is obtained by prolonging that side of the polygon which passes through the point; the rate will then depend on the steepness of this line, since, where the line is a side of the polygon, it represents the uniform motion which the body has during a certain interval. When the polygon looks like a curve the sides are very short, and any side, being prolonged both ways, will look like a tangent to the curve.

Now in considering the general case of varying motion we should have, instead of the above polygon which looks like a curve, an actual curve ; the difference between them being that, if we look at the curve-like polygon with a sufficiently strong microscope, we shall be able to see its angles, but however powerful a microscope we may apply to the curve it will always look like a curve. But there is this property in common, that if we draw a tangent to the curve at any point, then, since the steepness of this tangent will be exactly the same as the steepness of the curve at that particular point, it will give the rate for the motion represented by the curve, just as before the steepness of the prolonged small side of the polygon gave the rate for the motion represented by the polygon. That-is to say, the instantaneous velocity of a body in any position may be learnt from its curve of positions by drawing a tangent to this curve at the point corresponding to the position; for the steepness of this tangent will give us the velocity or rate which we want, since the tangent itself corresponds to a uniform motion of the same velocity as that belonging to the given varying motion at the particular instant. From this means of representing the rate we can see how it is that the instantaneous velocity of a body generally belongs to it only at an instant and not for any length of time however short; for the steepness of the curve is continually changing as we go from one part of it to another, and the curve is not straight for any portion of its length however small.

The problem of determining the instantaneous velocity in a given position is therefore reduced to the problem of drawing a tangent to a given curve. We have a sufficiently clear general notion of what is meant by each of these things, but the notion which is sufficient for purposes of ordinary discourse is not sufficient for the purposes of reasoning, and it must therefore be made exact. Just as we had to make our notion of the ratio of two quantities exact by means of a definition of the fourth proportional, or of the equality of two ratios which were expressed in terms of numbers, so here we shall have to make our idea of a velocity exact by expressing it in terms of measurable quantities which do not change.

We have no means of measuring the instantaneous velocity of a moving body; the only thing that we can measure is the space which it traverses in a given interval of time. In the case in which a body is moving uniformly, its instantaneous velocity, being always the same, is completely specified as soon as we know how far

R

the body has gone in a definite time. And, as we have already observed, the result is the same whatever this interval of time may be ; the rate of four miles an hour is the same as eight miles in two hours, or two miles in half an hour, or one mile in a quarter of an hour. But if a body be moving with a velocity which is continually changing, the knowledge of how far it has gone in a given interval of time tells us nothing about the instantaneous velocity for any position during that interval. To say, for instance, that a man has travelled a distance of four miles during an hour, does not give us any information about the actual rate at which he was going at any moment during the hour, unless we know that he has been going at a uniform rate. Still we are accustomed to say that in such a case he must have been going *on an average* at the rate of four miles an hour; and, as we shall find it useful to speak of this rate as an 'average velocity,' its general definition may be given as follows :—

If a body has gone over a certain distance in a certain time its *mean* or *average velocity* is that with which, if it travelled uniformly, it would get over the same distance in the same time.

This mean velocity is very simply represented by the help of the curve of positions. Let *a* and *b* be two points on the curve of positions; then the mean velocity between the position represented by *a* and that represented by *b* is given by the steepness of the straight line *a b*. This, moreover, enables us to make some progress towards a method of calculating instantaneous velocity, for we showed that the problem of finding the instantaneous velocity of a body is, in the above method of representation, the problem of drawing a tangent to a curve. Now the mean velocity of a body is defined in terms of quantities which we are already able to measure, for it requires the measurement of an interval of time and of the distance traversed during that interval; and further the *chord* of a curve, *i.e.* the line joining one point of it to another,

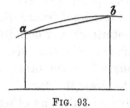

FIG. 93.

is a line which we are able to draw. If then we can find some means of passing from the chord of a curve to the tangent, the representation we have adopted will help us to pass from the mean to the instantaneous velocity.

§ 5. *On the Tangent to a Curve.*

Now let us suppose the chord *a b* joining the points on the curve to turn round the point *a*, which remains fixed; then *b* will travel along the curve

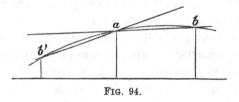

FIG. 94.

towards *a*; and if we suppose *b* not to stop in this motion until it has got beyond *a* to a point such as *b* on the other side, the chord will have turned round into the position *a b'*. Now, looking at the curve which

r 2

THE COMMON SENSE OF THE EXACT SCIENCES.

is drawn in the figure, we see that the tangent to the curve at a obviously lies between $a\,b$ and $b'\,a$. Thus if $a\,b$ turn round a so as to move into the position $a\,b'$ it will at some instant have to pass over the position of the tangent. At the instant when it passes over this position where is the point b? We can at once see from the figure that it cannot be anywhere else than at a, and yet we cannot attach any definite meaning to a line described as joining two coincident points. If we could, the determination of the tangent would be very easy, for in order to draw the tangent to the curve at a, we should merely say, Take any other point b on the curve; join $a\,b$ by a straight line; then make b travel along the curve towards a, and the position of the line $a\,b$ when b has got to a is that of the tangent at a. Here however arises the difficulty which we have already pointed out, namely, that we cannot form any distinct conception of a line joining two coincident points; two separate points are necessary in order to fix a straight line. But it is clear that, although it is not yet satisfactory, there is still something in the definition that is useful and correct; for if we make the chord turn from the position $a\,b$ to the position of the tangent at a, the point b does during this motion move along the curve up to the point a.

This difficulty was first cleared up and its explanation made a matter of common sense by Newton. The nature of his explanation is as follows:—Let us for simplicity take the curve to be a circle. If a straight stick be taken and bent so as to become part of a circle, the size of this circle will depend upon the amount of bending. The stick may be bent completely round until the ends meet, and then it will make a very small circle; or it may be bent very slightly indeed, and

then it will become part of a very large circle. Now,
conversely, suppose that we begin with a small circle,
and, holding it fast at one point, make it get larger
and larger, so that the piece we have hold of gets less
and less bent; then, as the circle becomes extremely
large, any small portion of it will more and more
nearly approximate to a straight line. Hence a circle
possesses this property, that the more it is magnified
the straighter it becomes; this property likewise belongs to all the curves which we require to consider.
It is sometimes expressed by saying that the curve is
straight in its elements, or in its smallest parts; but
the statement must be understood to mean only this,
that the smaller the piece of a curve is taken the
straighter it will look when magnified to a given
length.

Now let us apply this to the problem of determining
the position of a tangent. Let us suppose the tangent $a\,t$ of a circle to be already drawn, and that a

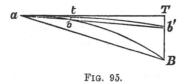

FIG. 95.

certain convenient length is marked off upon it; from
the end of this T let a perpendicular be drawn to meet
the circle in B, and let a be joined to B by a straight
line. We have now to consider the motion of the point
B along the circle as the chord $a\,B$ is turning round
a towards the position $a\,T$; and the difficulty in our way
is clearly that figures like $a\,B\,T$ get small, as for example $a\,b\,t$, and continue to decrease until they cease
to be large enough to be definitely observed. Newton

gets over this difficulty by supposing that the figure is always magnified to a definite size; so that instead of considering the smaller figure $a\,b\,t$ we magnify it throughout until $a\,t$ is equal to the original length $a\,T$. But the portion $a\,b$ of the circle with which we are now concerned is less than the former portion $a\,B$; consequently when it is magnified to the same length (or nearly so) it must appear straighter. That is to say, in the new figure $a\,b'\,T$, which is $a\,b\,t$ magnified, the point b will be nearer to the point T than B in the old one $a\,B\,T$; consequently, also, as b moves along to a the chord $a\,b$ will get nearer to the tangent $a\,T$, or, what is the same thing, the angle $t\,a\,b$ will get smaller. This last result is clear enough, because, as we previously supposed, the chord $a\,b$ is always turning round towards the position $a\,t$.

But now the important thing is that, by taking b near enough to a, we can make the curve in the magnified figure as straight as we please; that is to say, we

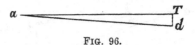

FIG. 96.

can make b' approach as near as we like to T. If we were to measure off from T perpendicularly to $a\,T$ any length, however small, say $T\,d$, then we can always draw a circle which shall have $a\,T$ for a tangent and which shall pass between T and d; and, further, if we like to draw a line $a\,d$ making a very small angle with $a\,T$, then it will still be possible to make b go so close to a that in the magnified figure the angle $b'\,a\,T$ shall be smaller than the angle $d\,a\,T$ which we have drawn.

Now mark what this process, which has been called

Newton's microscope, really means. While the figure
which we wish to study is getting smaller and smaller,
and finally disappears altogether, we suppose it to be
continually magnified, so as to retain a convenient size.
We have one point moving along a curve up towards
another point, and we want to consider what happens
to the line joining them when the two points approach
indefinitely near to one another. The result at which
we have arrived by means of our microscope is that,
by taking the points near enough together, the line
may be made to approach as near as we please to the
tangent to the curve at the point a. This, therefore,
gives us a definition of the tangent to a curve in
terms only of measurable quantities. If at a certain
point a of a curve there is a line $a\,t$ possessing the
property that by taking b near enough to a on the curve
the line $a\,b$ can be brought as near as we like to $a\,t$
(that is, the angle $b\,a\,t$ made less than any assigned
angle, however small), then $a\,t$ is called the tangent to
the curve at the point a. Observe that all the things
supposed to be done in this definition are things which
we know can be done. A very small angle can be
assigned; then, this angle being drawn, a position of
the point b can be found which is such that $a\,b$ makes
with $a\,t$ an angle smaller than this. A supposition is
here made in terms of quantities which we already
know and can measure. We only suppose in addition
that, however small the assigned angle may be, the
point b can always be found; and if this is possible,
then in the case in which the assigned angle is extremely small, the line $a\,b$ or $a\,t$ (for they now coincide) is called a tangent.

It is worth while to observe the likeness between
this definition and the one that we previously discussed

of the fourth proportional or of the equality of ratio. In that definition we supposed that, a certain fraction being assigned, if the first ratio were greater than this fraction, so also was the second ratio, and if less, less ; and the question whether these ratios were greater or less is one that can be settled by measurement and comparison. We then made the further supposition that whatever fraction were assigned the same result would hold good ; and we said that in that case the ratios were equal. Now in both of these definitions, applying respectively to tangents and to ratios, the difficulty is that we cause a particular supposition to be extended so as to be general ; for we assume that a statement which can be very easily tested and found true in any one case is true in an infinite number of cases in which it has not been tested. But although the test cannot be applied individually to all these cases in a practical way, yet, since it is true in any individual case, we know on rational grounds that it must be satisfied in general ; and therefore, justified by this knowledge, we are able to reason generally about the equality of ratios and about the tangents to curves.

Let us now translate the definition at which we have thus arrived from the language of curves and tangents into the language of instantaneous and mean velocities. The steepness of the chord of the curve of positions indicates the mean velocity, while the steepness of the tangent to the curve at any point indicates the instantaneous velocity at that point. The process of making the point *b* move nearer and nearer to the point *a* corresponds to taking for consideration a smaller and smaller interval of time after that moment at which the instantaneous velocity is wanted.

Suppose, then, the velocity of a body, viz. a railway

train, to be varying, and that we want to find what its value is at a given instant. We might get a very rough approximation to it, or in some cases no approximation at all, by taking the mean velocity during the hour which follows that instant. We should get a closer approximation by taking the mean velocity during the minute succeeding that instant, because the instantaneous velocity would have less time to change. A still closer approximation would be obtained were we to take the mean velocity during the succeeding second. In all motions we should have to consider that we could make the approximation as close as we like by taking a sufficiently small interval. That is to say, if we choose to name any very small velocity, such as one with which a body going uniformly would move only an inch in a century, then, by taking the interval small enough, it will be possible to make the mean velocity differ from the instantaneous velocity by less than this amount. Thus, finally, we shall have the following definition of instantaneous velocity : If there is a certain velocity to which the mean velocity during the interval succeeding a given instant can be made to approach as near as we like by taking the interval small enough, then that velocity is called the instantaneous velocity of the body at the given instant.

In this way then we have reduced the problem of finding the velocity of a moving body at any instant to the problem of drawing a tangent to its curve of positions at the corresponding point ; and what we have already proved amounts to saying that, if the position of the body be given in terms of the time by means of a curve, then the velocity of the body will be given in terms of the time by means of the tangent to this curve.

Now there are many curves to which we can draw tangents by simple geometrical methods, as, for example, to the ellipse and the parabola; so that, whenever the curve of positions of a body happens to be one of these, we are able to find by geometrical construction the velocity of the body at any instant. Thus in the case of a falling body the curve of positions is a parabola, and we might find by the known properties of the tangent to a parabola that the velocity in this case is proportional to the time. But in the great majority of cases the problem of drawing a tangent to the curve of positions is just as difficult as the original problem of determining the velocity of a moving body, and in fact we do in many cases solve the former by means of the latter.[1]

§ 6. On the Determination of Variable Velocity.

What is actually wanted in every case will be apparent from the consideration of the problem we have just mentioned—that of a body falling down straight. We note, from the experience of Galilei, that the whole distance which the body has fallen from rest at any instant is proportional to the square of the time ; in fact, to obtain this distance in feet we must multiply the number of seconds by itself and the result by a number a little greater than sixteen. Thus, for instance, in five seconds the body will have fallen rather more than twenty-five times sixteen feet, or 400 feet. Now what we want is some direct process of proving that when the distance traversed is proportional to the square of the time the velocity is always proportional to the time. In the present case we can find the velocity at the end of a given number of seconds by multiplying that number by thirty-two feet ; thus at the end of five seconds the velocity of the body will be 160 feet per

---

[1] [The method is due to *Roberval* (1602-1675).—K. P.]

second.[1] Now as a matter of fact a process (of which there is a simple example in the footnote) has been worked out, by which from any algebraical rule telling us how to calculate the distance traversed in terms of the time we can find another algebraical rule which will tell us how to calculate the velocity in terms of the time. One case of the process is this : If the distance traversed is at any instant a times the nth power of the time, then the velocity at any instant will be na times

[1] The following may be taken as a proof. Let a be the distance from rest moved over by the body in t' seconds, b that moved over by it in $t + t'$ seconds, so that t' seconds is the interval we take to find out the mean velocity. Now by our rule just quoted, since a feet are passed over in t seconds, we have

$$a = 16t^2,$$

and similarly $\qquad b = 16(t + t')^2 = 16 \ (t^2 + 2tt' + t'^2).$

Hence we have $\quad b - a = 16(t^2 + 2tt' + t'^2) - 16t^2$
$$= 16(2tt' + t'^2)$$
$$= 16t'(2t + t'),$$

giving the distance moved over in the interval t'. But the mean velocity during this interval is obtained by dividing the distance moved over by the time taken to traverse it ; hence the mean velocity in our case for the interval of t' seconds immediately succeeding the t seconds

$$= \frac{b - a}{t'}$$
$$= \frac{16t'(2t + t')}{t'}$$
$$= 16(2t + t')$$
$$= 32t + 16t'.$$

Now if we look at this result, which we have obtained for the mean velocity, we see that there are two terms in it. The first, viz. $32t$, is quite independent of the interval t' which we have taken ; the second, viz. $16t'$, depends directly on it, and will therefore change when we change the interval. Now the distance per second represented by $16t'$ feet can be made as small as we like by taking t' small enough ; so that the mean velocity during the interval t' seconds succeeding the given instant can be made to approach $32t$ feet per second as near as we like by taking t' small enough. Recurring to our definition of instantaneous velocity, it is now evident that the instantaneous velocity of our falling body at the end of t seconds is $32t$ feet per second.

the $(n-1)$th power of the time. It is by means of this process of altering one algebraical rule so as to get another from it that both of the problems which we have shown to be equivalent to one another are solved in practice.

There is yet another problem of very great importance in the study of natural phenomena which can be made to depend on these two. When a point moves along a straight line the distance of it from some fixed point in the line is a quantity which varies from time to time. The rate of change of this distance is the same thing as the velocity of the moving point; and the rate of change of any continuous quantity can only be properly represented by means of the velocity of a point.

Thus, for instance, the height of the tide at a given port will vary from time to time during the day, and it may be indicated by a mark which goes up and down on a stick. The rate at which the height of the tide varies will obviously be the same thing as the velocity with which this mark goes up and down. Again the pressure of the atmosphere is indicated by means of the height of a mercury barometer. The rate at which this pressure changes is obviously the same thing as the velocity with which the surface of the mercury moves up and down. Now whenever we want to describe the changes which take place in any quantity in terms of the time, we may indeed roughly and approximately do so by means of a table. But this is also the most troublesome way; the proper way of describing them is by drawing a curve in which the *abscissa*, or horizontal distance, at any point represents the time, while the height of the curve at that point represents the value of the quantity at that time (see p. 184). Whenever this is done we practically suppose the variation of the quantity to be represented by the motion of the point on a curve. The quantity can only be adequately represented by marking off a length proportional to it on a line; so that if the quantity varies then the length marked off will vary, and consequently the end of this length will move along the curve. The rate at which the quantity varies is the rate at which this point moves; and when the values of the quantity for different times are represented by the perpendicular distances of points on a curve from the line which represents the time, its rate of variation is determined by the tangent to that curve.

§ 7. *On the Method of Fluxions.*

Hence we have three problems which are practically the same. First, to find the velocity of a moving point when we know where it is at every instant; secondly, to draw a tangent to a curve at any point; thirdly, to find the rate of change of a quantity when we know how great it is at every instant. And the solution of them all depends upon that process by which, when we take the algebraical rule for finding the quantity in terms of the time, we deduce from it another rule for finding its rate of change in terms of the time.

This particular process of deriving one algebraical rule from another was first investigated by Newton. He was accustomed to describe a varying quantity as a *fluent,* and its rate of change he called the *fluxion* of the quantity. On account of these names, the entire method of solving these problems by means of the process of deriving one algebraical rule from another was termed the *Method of Fluxions.*

In general the rate of variation of a quantity will

itself change from time to time; but if we consider
only an interval very small as compared with that required for a considerable variation of the quantity, we
may legitimately suppose that it has not altered much
during that interval. This is practically equivalent to
supposing that the law of change has been uniformly
true during that interval, and that the rate of change
does not differ very much from its mean value. Now
the mean rate of change of a quantity during an interval
of time is just the difference between the values of the
quantity at the beginning and at the end divided by
the interval. If any quantity increased by one inch in
a second, then, although it may not have been increasing uniformly, or even been increasing at all during the
whole of that second, yet during the second its mean
rate of increase was one inch per second. Now if the
rate of increase only changes slowly we may, as an
approximation, fairly suppose it to be constant during
the second, and therefore to be equal to the mean rate;
and, as we know, the smaller the interval of time is, the
less is the error arising from this supposition. This is,
as a matter of fact, the way in which that process is
established by means of which a rule for calculating
position is altered into a rule for calculating velocity.
The difference between the distances of the moving
point from some fixed point on the line at two different
times is divided by the interval between the times, and
this gives the mean rate of change during that interval.
If we find that, by making the interval smaller and
smaller, this mean rate of change gets nearer and
nearer to a certain value, then we conclude that this
value is the actual rate of change when we suppose the
interval to shrink up into an instant, or that it is, as
we call it, the instantaneous rate of change.

Because two differences are used in the argument which establishes the process for changing the one rule into the other, this process was called, first in other countries and then also in England, the *Differential Calculus*. The name is an unfortunate one, because the rate of change which is therein calculated has nothing to do with differences, the only connection with differences being that they are mentioned in the argument which is used to establish the process. However this may be, the object of the differential calculus or of the method of fluxions (whichever name we choose to give it) is to find a rule for calculating the rate of change of a quantity when we have a rule for calculating the quantity itself; and we have seen that when this can be done the problem of drawing a tangent to a curve and that of finding the velocity of a moving point are also solved.

§ 8. *Of the Relationship of Quantities, or Functions.*

But we not only have rules for calculating the value of a quantity at any time, but also rules for calculating the value of one quantity in terms of another quite independently of the.time. Of the former class of rules an example is the one mentioned above for calculating the rise of the tide. We may either write down a formula which will enable us to calculate it at a given instant, or we may draw a curve which shall represent its rise at different times of the day. Of the second kind of rule a good example is that in which the pressure of a given quantity of gas is given in terms of its volume when the temperature is supposed to be constant; the algebraical statement of the rule giving the relation between them is that the two things vary inversely as one another, or that the product representing them is constant. Thus if we compress a mass of air to one-half of its natural volume the pressure will become twice as great, or will be, as it is called, two 'atmospheres.' And so if we compress it to one-fifth of the volume the pressure will become five times as great, or five atmospheres.

If we like to represent this by a figure we shall draw a curve in which the abscissa, or horizontal distance from the starting point, will represent the volume, and a vertical line drawn through the extremity of this abscissa will represent the pressure. For any particular temperature the curve traced out by the extremity of the line representing the pressure will be a hyperbola having one asymptote vertical and the other horizontal; and for different temperatures we shall have different hyperbolas with the same asymptotes. Thus every point in the plane will represent a particular state of the body, since some hyperbola can be drawn through it; the horizontal distance of the point from the origin will represent the volume, and its vertical distance the pressure, while the particular hyperbola on which it lies will indicate the temperature. We have here an example of the physical importance of a *family of curves*, to which reference was made in the preceding chapter (see p. 163).

When the connection between two quantities has to be found out by actual observation, this is done by properly plotting down points on paper (as in § 11, Chap. IV.) to represent successive observations. Thus in the case of air the pressure would be observed for different values of the volume. For each of these observed pairs of values a point would be marked in the plane; and when a sufficient number had been marked it would become obvious to the eye that, roughly speaking, the point lay on a hyperbolic curve. But it is to be noticed that it is only roughly that this result holds, because observations are never so accurate that the curve does not require to be drawn pretty freely in passing through the points. But directly the geometer has seen that the shape of the curve is hyperbolic he recognises the law that pressure varies inversely as volume.

We have here the relation between two quantities expressed by means of a curve. Whenever two quantities are related in some such way, so that one of them being given the other can be calculated or found, each is said to be a *function* of the other. Now a function may be supposed to be given either by an algebraical rule or by a curve. Thus to find the pressure corresponding to a given volume we might say that a certain number was to be divided by the number representing the volume, and the result would be the number of units of pressure; or we might say that from the given point of the horizontal line which represented the volume a perpendicular was to be drawn and continued till it met the curve, and that the ordinate (or the part of this between the horizontal line and the curve) represented the pressure. We have thus a connection established between the science of geometry and the science of quantity, as, for example, the relation between the two quantities, volume and pressure, is expressed by means of a certain curve.

Now every connection between two sciences is a help to both of them. When such a connection is established we may both use the known theorems about quantities in order to investigate the nature of curves (and this is, in fact, the method of co-ordinates introduced by Descartes), or we may make use of

s

known geometrical properties of curves in order to find out theorems about the way in which quantities depend upon one another. For the first purpose the relation between the two quantities is regarded as an equation. Thus, instead of saying that a pressure varies inversely as a volume we should prefer to say that the product of the pressure and the volume is equal to a certain constant, the temperature being supposed unaltered; or, paying attention only to the geometrical way of expressing this, we should say that, for points along the curve we are considering, the product of the abscissa and the ordinate is equal to a certain fixed quantity. This is written for shortness

$$xy = c^2,$$

and from such an equation all the properties of a hyperbola may be deduced.

But we may also make use of the properties of known curves in order to study the ways in which quantities can depend on one another. Thus the per-

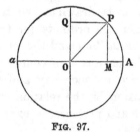

FIG. 97.

pendicular distance PM from the point P of the circle to a fixed diameter AOa is a quantity whose ratio to the radius OP depends in a certain definite way upon the magnitude of the angle POA, or, what is the same thing (p. 143), upon the length of the arc AP. The ratio is in fact what we have termed the sine of the angle, or,

as it is sometimes called, the sine of the arc. If the arc
AP is made proportional to the time, or, what is the
same thing, if P is made to move uniformly round the
circle, then the length of the line PM will represent
the distance from the centre O of a point Q oscillating
according to a law which is defined by this geometrical
construction. This particular kind of oscillation, which
is called *simple harmonic motion*, occurs when the air
is agitated by sound, or the ether by light, or when
any elastic body is set into a tremor. Relations such
as that which we have just mentioned between arcs of
a circle and straight lines drawn according to some
simple constructions in the circle give rise to what are
often termed *circular* functions. Thus the trigonometrical ratios considered in § 7 of Chapter IV. are
functions of this kind. We have also *hyperbolic* functions, depending on the hyperbola in somewhat the
same way in which circular functions depend upon the
circle, and *elliptic* functions, so called because by means
of them the length of the arc of an ellipse can be calculated.

But the most valuable method of studying the
properties of functions is derived from the considerations of which we have been treating in this chapter,
viz. considerations of the rate of change of quantities.
When the relation between two quantities is known, the
relation between their rates of change can be found by
a known algebraical process; and we have shown that
the problem of finding this relation ultimately comes
to the same thing as the problem of drawing a tangent
to the curve which expresses the relation between the
two original quantities. Thus, in the case we previously considered of two quantities whose product is
constant or which vary inversely as one another, it is

clear that one must increase when the other decreases ; it is found that the ratio of these rates of change is equal to the ratio of the quantities themselves. Thus the rate of change of the volume of a gas is to the rate of change of the pressure (the temperature being kept constant) as the volume is to the pressure, it being always remembered that an increase of the one implies a decrease of the other.

The consideration of this ratio of the rates of change is of great importance in determining one of the fundamental changeable properties of a body, namely, its *elasticity*. We define the elasticity of a gas as the change of pressure which will produce a given *contraction*; where by the term contraction is meant the change in the volume divided by the whole volume before change. Thus if the volume of a gas diminished one per cent., it would experience a contraction of $\frac{1}{100}$th. If then, in accordance with our definition, we divide the pressure necessary to produce this contraction by $\frac{1}{100}$, or, what is the same thing, multiply it by 100, we shall get what is called the elasticity. Now in our case the change of pressure divided by the whole pressure is equal to what we have called the contraction, that is, to $\frac{1}{100}$; and therefore the change of pressure is equal to $\frac{1}{100}$th of the whole pressure. But we have just proved that the elasticity is 100 times the change of pressure necessary to produce the contraction we have been considering, and it is therefore equal to the whole pressure. Consequently the elasticity of a gas is measured by the pressure of the gas.

§ 9. *Of Acceleration and the Hodograph.*

We may then consider the rate of change of any measurable quantity as another quantity which we can find; and we have derived our notion of it from the velocity of a moving point. In the simplest case, when this point is moving along a straight line, the rate at which it is going is the rate of change of its distance from a point fixed in the line. But in the general case, when the point is moving not on a straight line, but along any sort of curve, we shall not give a complete description of its state of motion if we only say how fast it is going; it will be necessary to say in addition in what direction it is going. Hence we must not only measure the quantity of a velocity, but also a certain quality of it, viz. the direction. Now we do as a matter of fact contrive to study these two things together, and the method by which we do so is perhaps one of the most powerful instruments by which the scope of the exact sciences has been extended in recent times. Defining the velocity of a moving point as the rate of change of its position, we are met by the question, What is its position?

This question has been answered in the preceding chapter. The position of a moving point is determined when we know the directed step or vector which connects it with a fixed point. If then the velocity of the moving point means the rate of change of its position, and if this position is determined by the vector which would carry us from some fixed point to the moving point, in order to understand velocity we shall have to get a clear conception of what is meant by the rate of change of a vector.

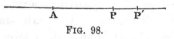

FIG. 98.

Let us go back for a moment to the simpler case of a point moving along a straight line; its position is determined by means of the step AP from the point A fixed in the straight line to the moving point P. Now this step alters with the motion of the point; so that if the point comes to P' the step is changed from AP to AP'. How is this change made in the step? Clearly by adding to the original step AP the new step PP', and we specify the velocity of P by saying at what rate this addition is made.

Now let us resume the general case. We have the fixed point A given; and the position of the moving point P is determined by means of the step AP. As P moves about, this step gets altered, so that when P comes to P' this step is AP'; it is therefore obvious that it is altered not only in magnitude but also in direction. Now the change may be made by adding to the original step AP the new step PP'; and it is quite clear that if we go from A to P and then from P to P' the result is exactly the same as if we had gone

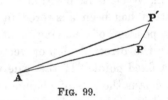

FIG. 99.

directly from A to P'. The question then is: At what rate does this addition take place, or what step per second is added to the position? The answer as before is of the nature of a step or vector—that is, the change of position of the moving point has not only magnitude but direction. We shall therefore have to say that the rate of change of a step or vector is always so many feet per second in a certain direction.

To sum up, then, we state that the velocity of a

moving point is the rate of change of the step which specifies the position; and that in order to describe accurately this velocity, we must draw a line of given length in a given direction; we observe also that the rate of change of a directed quantity is itself a directed quantity. This last remark is of the utmost importance, and we shall now apply it to a consideration of the velocity itself.

If a point is moving uniformly in a straight line its velocity is always the same in magnitude and the same in direction; and consequently a line drawn to represent it would be unaltered during the motion. But if a point moves uniformly round a circle its velocity, although always the same in magnitude, will be constantly changing in direction, and the line which specifies this velocity will thus be always of the same length, but constantly turning round so as always to keep parallel with the direction of motion of the moving point. And so, generally, when a point is moving along any kind of curve let us suppose that through some other point, which is kept fixed, a line is always drawn which represents the velocity of the moving point both in magnitude and direction. Since the velocity of the moving point will in general change, this line will also change both in size and in direction, and the end of it will trace out some sort of curve. Thus in the case of the uniform circular motion, since the velocity remains constant, it is clear that the end of the line representing the velocity will trace out a circle; in the case of a body thrown into the air the end of the corresponding line would be found to describe a vertical straight line. This curve described by the end of the line which represents the velocity at any instant may be regarded as a map of the motion, and was for that reason called by Hamilton the *hodograph*. If we know the path of the moving point and also the hodograph of the motion, we can find the velocity of the moving point at any particular position in its path. All we have to do is to draw through the centre of reference of the hodograph a line parallel to the tangent to the path at the given position; the length of this line will give the rate of motion, or the velocity of the point as it passes through that position in its path. Hamilton proved that in the case of the planetary orbits described about the sun the hodograph is always a circle. In this case it possesses other interesting properties, as, for example, that the amount of light and heat received by the planet during a given interval of time is proportional to the length of the arc of the hodograph between the two points corresponding to the beginning and end of that interval.

But the great use of the hodograph is to give us a clear conception of the rate of change of the velocity. This rate of change is called the *acceleration*. Now, it must not be supposed that acceleration always means an *increase* of velocity, for in this case, as in many others, mathematicians have adopted for use one word to denote a change that may have many directions; thus a decrease of velocity is called a negative acceleration. This mode of speaking, although rather puzzling at first, becomes a help instead of a confusion when one is accustomed to it. Now a velocity may be changed in magnitude without altering its direction— that is to say, it may be changed by adding it to a velocity parallel to itself. In this case we say that the acceleration is in the direction of motion. But a velocity may also be changed in direction without being changed in magnitude, and we have seen that then the hodograph is a circle. The velocity is altered by adding to it a velocity perpendicular to itself, for the tangent at any point to a circle is at right angles to the radius drawn to that point, and in this case we may say that the acceleration is at right angles to the direction of motion. But in general both the magnitude and the direction of the velocity will vary, and then we shall see that the acceleration is neither in the direction of motion nor at right angles to it, but that it is in some intermediate direction.

If we consider the motion in the hodograph of the end of the line representing the velocity, we observe the motion of a point whose position is defined by the step to it from the centre of the hodograph. Now this step is just the velocity of the point P in the original curve, for the line OQ is supposed to be drawn at every instant

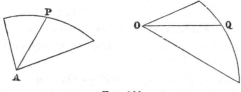

FIG. 100.

to represent the velocity of P in magnitude and direction. Now we saw that the rate of change of the step from some fixed point A to P was the velocity of P. Hence, since the step OQ drawn from the fixed point O to Q defines the position of Q, it is obvious that the rate of change of the step OQ is the velocity of Q. Since OQ represents the velocity of P, it follows that the velocity of the point Q describing the hodograph is the rate of change of the velocity of P; that is to say, it is the acceleration of the motion of P. This acceleration being the velocity of Q, and a velocity being as we have seen a vector, it at once follows that the acceleration is a vector or directed quantity.

In changing the magnitude and direction of the velocity of a moving point we may consider that we are pouring in, as it were, velocity of a certain kind at a certain rate. In the case of a stone thrown up obliquely and allowed to fall again the path described is a parabola, and the direction of motion, which originally pointed obliquely upwards, turns round and becomes horizontal, and then gradually points more and more downwards. But what has really been happening the whole time is that velocity straight downwards has been continually added at a uniform rate during every second, so that the original velocity of the stone is compounded with a velocity vertically downwaros, increasing uniformly at the rate of thirty-two feet a second. In this case, then, we say that the acceleration, or rate of change per second of the velocity of the stone, is constant and equal to thirty-two feet a second vertically downwards.

If we whirl anything round at the end of a string we shall be continually pouring in velocity directed towards the end of the string which is held in the hand; and since the velocity of the body which is being whirled is perpendicular to the direction of the string, the added velocity is always perpendicular to the existing velocity of the body. And so also when a planet is travelling round the sun there is a continual pouring in of velocity towards the sun, or, as we say, the acceleration is always in the line joining the planet to the sun. In addition it is in this case found to vary inversely as the square of the distance from the sun.

§ 10. *On the Laws of Motion.*

These examples prepare us to understand that law of motion which is the basis of all exact treatment of physics. When a body is moving let us consider what it is that depends upon the circumstances, meaning by the ' circumstances ' the instantaneous position relative to it of other bodies as well as the instantaneous state of the body itself irrespective of its motion. We might at first be inclined to say that the velocity of the body depends on the circumstances, but very little reflection will show us that in the same circumstances a body may be moving with very different velocities. At a given height above the earth's surface, for example, a stone may be moving upwards or downwards, or horizontally, or at any inclination, and in any of these modes with any velocity whatever; and there is nothing contrary to nature in supposing a motion of this sort. Yet we should find that, no matter in what way the stone may move through a given position, the rate of change per second of its velocity will always be the same, viz. it will be thirty-two feet per second vertically downwards. When we push a chair along the ice, in order to describe the circumstances we must state the compression of those muscles which keep our hands against the chair. Now the rate at which the chair moves does not depend simply upon this compression; for a given amount of push may be either starting the chair from rest or may be quickening it when it is going slowly, or may be keeping it up at a high rate.

What is it, then, which does depend upon the circumstances ? In whichever of these ways, or in whatever other way this given amount of push is used, its result in every case is obviously to change the rate of

motion of the chair; and this change of the rate of
motion will vary with the amount of push. Hence it is
the rate of change of the velocity, or the acceleration of
the chair which depends upon the circumstances, and
these circumstances are partly the compression of our
muscles and partly the friction of the ice; the one is
increasing and the other is diminishing the velocity in
the direction in which the chair is going.

The law of motion to which allusion has just been
made is this:—The acceleration of a body, or the rate of
change of its velocity depends at any moment upon the
position relative to it of the surrounding bodies, but
not upon the rate at which the body itself is going.
There are two different ways in which this dependence
takes place. In some cases, as when a hand is pushing
a chair, the rate of change of the velocity depends on
the state of compression of the bodies in contact; in
other cases, as in the motion of the planets about the
sun, the acceleration depends on the relative position
of bodies at a distance.

The acceleration produced in a body by a particular
set of surrounding circumstances must in each case be
determined by experiment, but we have learnt by experience a general law which much simplifies the experiments which it is necessary to make.   This law is as
follows:—If the presence of one body alone produces a
certain acceleration in the motion of a given body, and
the presence of a second body alone another acceleration; then, if both bodies are present at the same time,
the one has in general no effect upon the acceleration
produced by the other. That is, the total acceleration of the moving body will be the combination of the
two simple accelerations; or, since accelerations are
directed quantities, we have only to combine the simple

accelerations, as we did vector steps in § 3 of the preceding chapter, in order to find the result of superposing two sets of surrounding circumstances.

Now while this great law of nature simplifies extremely our consideration of the motion of the *same* body under different surrounding circumstances, it does not enable us to state anything as to the motion of *different* bodies under the same surrounding circumstances. This case, however, is amply provided for by another comprehensive law which experience also has taught us. We may thus state this third all-important law of motion :—The ratio of the accelerations which any two bodies produce in each other by their mutual influence is a constant quantity, quite independent of the exact physical characteristics of that influence. That is to say, however the two bodies influence one another, whether they touch or are connected by a thread or being at a distance still alter one another's velocities, this ratio will remain in these and all other cases the same.

§ 11. *Of Mass and Force.*

Let us see how we can apply this law. Suppose we take some standard body P and any other Q, and note the ratios of the accelerations they produce in each other under any of the simplest possible circumstances of mutual influence. Let the ratio determined by experiment be represented by m, or m expresses the ratio of the acceleration of the standard body P to that of the second body Q. This quantity m is termed the *mass* of the body Q. Let m' be the ratio of the accelerations produced in the standard body P and a third body R by their mutual influence. Now the law as it stands above enables us to treat only of the ratio of the accelerations of *P* and *Q*, or again of *P* and *R* under varied circumstances of mutual influence. It does not tell us anything about the ratio of the accelerations which *Q* and *R* might produce in each other. Experience, however, again helps us out of our difficulties and tells us that if *Q* and *R* mutually influence each other, the ratio of the acceleration of *Q* to that of *R* will be *inversely* as the ratio of *m* to *m'*. If then we choose to term unity the *mass* of our standard body, we may state generally that *mutual accelerations are inversely as masses*. Hence, when we have once determined the masses of bodies we are able to apply our knowledge of the effect of any set of circumstances on *one* body, to calculate the effect which the same circumstances would produce upon any other body.

The reader will remark that mass as defined above is a ratio of accelerations, or in other words a mere numerical constant experimentally deducible for any two bodies. It is found that for two bodies of the same uniform substance, their masses are proportional to their volumes. This relation of mass to volume has given rise to much obscurity. An indescribable something termed *matter* has been associated with bodies. Bodies are supposed to consist of matter filling space, and the mass of a body is defined as the amount of matter in it. An additional conception termed *force* has been introduced and is supposed to be in some way resident in matter. The force of a body *P* on a body *Q* of mass *m* is a quantity proportional to the mass *m* of *Q* and to the acceleration which the presence of *P* produces in the motion of *Q*. It will be obvious to the reader that this conception of force no more explains why the presence of *P* tends to change the velocity of *Q*, than the conception of matter explains why mutual accelerations are inversely as masses. The custom of basing our ideas of motion on these terms ' matter ' and ' force ' has too often led to obscurity, not only in mathematical, but in philosophical reasoning. We do not know *why* the presence of one body tends to change the velocity of another ; to say that it arises from the force resident in the first body acting upon the matter of the moving body is only to slur over our ignorance. All that we do know is that the presence of one body may tend to change the velocity of another, and that, if it does, the change can be ascertained from experiment, and obeys the above laws.

To calculate by means of the laws of motion from the observed effects on a simple body of a simple set of circumstances the more complex effects of any combination of circumstances on a complex body or system of bodies is the special function of that branch of the exact sciences which is termed *Applied Mathematics.*

LONDON : PRINTED BY
SPOTTISWOODE AND CO., NEW-STREET SQUARE
AND PARLIAMENT STREET

Printed in the United States
By Bookmasters